SCIENCE ACTIVITIES

Environment and Living Organisms

Paul Spychal

Hodder & Stoughton

LONDON SYDNEY AUCKLAND TORONTO

About the book

This book arose as a result of the author's concern at the dearth of material suitable for teaching balanced GCSE science. The text is presented as an integrated series of activities each of which incorporates a main body together with a follow up section. Group discussion work is emphasised throughout. The questions attempt to develop student skills by encouraging them to think about science and its relevance to their everyday lives.

Dedication: For Peter, Carl and Adam

Acknowledgements

The publisher would like to thank the following for their permission to reproduce copyright material:

British Gas plc for the extracts from their British Gas Safety Leaflet; The British Medical Journal for 'The case for compulsory pasteurisation' by James Howie published in the *British Medical Journal* 17 August 1985; N.S. Galbraith for an adapted version on p63 of the table from 'Investigation and control of an acute episode of disease.' *Medicine International* 1984; 2:1–7; Penguin Books for 'Cousin Lesley's See-Through Stomach' from *Gargling with Jelly* by Brian Patten (Kestrel Books, 1985) copyright © Brian Patten; Sidgwick & Jackson for the table on page 29 from *The Yearbook of Astronomy* by Patrick Moore.

The author and publishers would like to thank the following for permission to use their photographs:

National Dairy Council p5; Science Museum p7; Associated Press Photo p11; J. Sommer/The Danish Tourist Board p14; Museum of Antiquities of the University and Society of Antiquaries of Newcastle upon Tyne p15; U.S. Geological Survey Dept. of the Interior p24; NASA p28; NASA/Science Photo Library p30; Barnaby's Picture Library p50 (top), p61 (top); Bubbles Photo Library p50 (centre and bottom); Tony Craddock/Science Photo Library p55 (top); P. Morris/Ardea Photographics p55 (bottom); Science Photo Library pp56, 61 (bottom); Sally Anne Thompson Animal Photography p59; Trebor Snook/ICCE p63.

© 1990 Paul Spychal

First published in Great Britain 1990

British Library Cataloguing in Publication Data

Spychal, Paul
  Science activities: environment and living organisms
  1. Science. For schools
  I. Title
  500

ISBN 0 340 49919 2

Typeset in Ehrhardt by Gecko Ltd, Bicester, Oxon
Printed and bound in Hong Kong for the educational publishing division of Hodder and Stoughton Limited, Mill Road, Dunton Green, Sevenoaks, Kent by Colorcraft Ltd

ENVIRONMENT AND LIVING ORGANISMS

Contents

List of Attainment Targets

1	Exploration of Science	10	Forces
2	The Variety of Life	11	Electricity and Magnetism
3	Processes of Life	12	IT and Microelectronics
4	Genetics and Evolution	13	Energy
5	Human Influences on the Earth	14	Sound and Music
6	Types and Uses of Materials	15	Using Light and e/m Radiation
7	Making New Materials	16	The Earth in Space
8	Explaining how Materials Behave	17	The Nature of Science
9	Earth and Atmosphere		

Science activist

This book makes you think about science. *Science Activities* tries to develop your science skills. By using the book you will become a *science activist*. You will be asked to:

Discuss

Have a chat and have a natter
Talk about the problem posed.
Don't argue or upset the teacher
Just make sure your mind's not closed!

Find out

What on Earth is a peat bog?
How on Earth do I find out?
What is meant by the "mass media"?
I might try an encyclopaedia!

Be mathematical

One and one make two I think
Although I cannot prove it.
Graphs to plot and to interpret
That's the way to do it.

Devise experiments

Think, think, think, think.
Worry, worry, worry, worry.
Plan, plan, plan, plan.
Musn't do this in a hurry!

Draw

Pencil, paper, steady hand
Neatly draw and neatly label.
Subject's big so draw to scale
Then you won't fall from the table!

Design

Designer jeans, designer loo
It seems to be the rage.
I bet I could design as well
Just give me a blank page!

Write

ABC please help me
To write this corny letter.
XYZ then to bed
Could you have done it better?

Understand passages

What's this all about then Flo?
I haven't got a clue.
If you give it one more go
You'll know just what to do!

Pasteurisation

Wotta lotta bottle

Heat exchanger

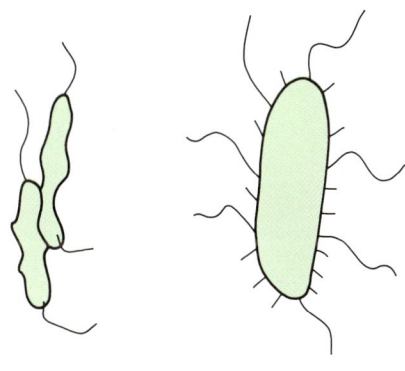

Campylobacter   Salmonella

Fig. 1

Bacteria which may contaminate food

Have you ever drunk milk that has not been treated or pasteurised? Pasteurisation involves heating milk to 72°C for at least fifteen seconds. It is then cooled quickly and bottled. The process of pasteurisation kills harmful bacteria. *Salmonella* and *Campylobacter*, fig. 1, are bacteria which may contaminate food and lead to food poisoning. Have you ever had food poisoning? It is not a pleasant experience. **Enteritis** is the inflammation of the intestine which can be caused by bacteria. This leads to food poisoning with the symptoms of diarrhoea (the runs), vomiting and abdominal pain. Correct pasteurisation should make milk harmless. This means you suffer less from the runs!

Carefully read the article on the case for compulsory pasteurisation.

The case for compulsory pasteurisation

In August 1983 compulsory pasteurisation of milk was introduced in Scotland, leading to a fall in milk borne infections. Meanwhile, in England and Wales there was no change in legislation and no fall in the number of infections. This amounts to a convincing case for compulsory pasteurisation throughout Britain.

Milk borne **salmonellosis** was particularly serious in Scotland during the 1970s and early 1980s, when 50 outbreaks affected at least 3 518 people and caused 12 deaths. In addition, there were several major outbreaks of campylobacter infection. The costs of one salmonella outbreak were estimated to lie between £236 000 and £3 222 000 and the mid-range average cost for each patient was £2637. But after compulsory pasteurisation, during 1983–4, eight outbreaks of milk borne salmonellosis affected only 46 people – all in dairy farming communities. No outbreaks were reported from the general community, whereas 14 general outbreaks had affected over 1090 people in the previous three years.

In 1983 two household episodes of campylobacter enteritis were attributed to raw milk. One of these happened before compulsory pasteurisation, and the other was in a family who had returned from a holiday in England. In 1984 there were no outbreaks of campylobacter enteritis.

Only temporary exemptions from pasteurisation were granted to a few small dairy farms on outer islands, which produce less than 0.02 per cent of Scottish milk. Concern remains only over dairy farm staff and their families being given free raw milk.

In contrast to Scotland, in England and Wales during 1983–4 some 20 outbreaks of milk borne salmonellosis affected at least 518 people.

Twelve of these outbreaks occurred in the general community in people who had bought raw milk direct from dairy farms. In the same years there were 6 proved outbreaks of campylobacter enteritis and 10 other suspected outbreaks without sufficient microbiological and epidemiological evidence to prove the diagnosis.

Although less than 3 per cent of the 6 000 million litres of milk sold annually in England and Wales is untreated, half of this is produced in the north west, where 70 per cent of the 1983–4 outbreaks occurred. Legislation is expected in November 1985 that will reduce the availability of untreated milk from shops – but it will not forbid direct 'farm gate' sales or local retail deliveries of farm bottled raw milk. At present legislation covers only cows' milk, but the milk of goats and sheep should be included as soon as possible. To reduce still further the risks of milk borne infection, farm water supplies need to be better controlled and pasteurisation plants must be well maintained, thoroughly cleaned and efficiently operated. Failures are sometimes due to insufficient pasteurisation or to contamination of pasteurised milk by raw milk or dirty containers. But these are minor matters compared with the need for compulsory pasteurisation throughout Britain of all animal milk intended for human consumption.

Passionate opponents of compulsory pasteurisation are likely to remain unconvinced for ever since they rely on feelings rather than evidence. Often these arguments seem to rest on a desire not to go against what nature intended. It might be argued that nature never intended man to drink animals' milk anyway, but these arguments count for little in the face of continuing milk borne infections being a costly nuisance and sometimes leading to death.

James Howie
British Medical Journal 17 August 1985

1 Copy and complete the following sentences.
   a   Examples of bacteria which cause food poisoning are . . . . and . . . .
   b   . . . . , ... and . . . . are symptoms of food poisoning.
   c   . . . . is the inflammation of the intestine.
   d   The process of making milk harmless is called . . . . .

2 a   When was compulsory pasteurisation of milk introduced in Scotland?
   b   How many outbreaks of milk borne salmonellosis took place, in England and Wales, during 1983–4?

3 a   How is the pasteurisation of milk carried out?
   b   Why is milk pasteurised?
   c   How can pasteurisation plants be sure to reduce still further the risks of milk borne infection?

4 a   What percentage of milk sold, in England and Wales, is untreated?
   b   What total volume of milk is sold annually in England and Wales?
   c   Calculate the volume of untreated milk sold in England and Wales.

5 a   How many people, in Scotland, were affected by milk borne salmonellosis in the 1970s and early 1980s?
   b   What was the mid-range average cost, in Scotland, for each patient with *salmonella*?
   c   Calculate the total cost, to the National Health Service, of *salmonella* outbreaks in the 1970s and early 1980s in Scotland.
   d   Do you think pasteurising milk will save the country money?

6 Answer this question in groups of four students. Discuss and decide, as a group, whether pasteurisation should be made compulsory or not. List the arguments for and against compulsory pasteurisation.

7 Write a letter to your Member of Parliament or your local newspaper showing either:

● your opposition to compulsory pasteurisation of milk or

● your concern that unpasteurised milk is still available in England and Wales.

Milk matters

LOUIS PASTEUR

Louis Pasteur

1 Find out about Louis Pasteur, who lived between 1822 and 1895. Write short notes about his important discoveries.

2 a   How much milk is sold, daily, in England and Wales?
   b   Calculate how much milk is consumed, daily, by the average person in England and Wales. Assume a population of 50 million.
   c   Is milk good for you? Find out if full fat milk is healthier than skimmed milk.
   d   Should the average person drink more or less milk than they do at present? Explain your answer.

3 What evidence is there, in the article, to suggest that untreated milk can cause food poisoning?

4 a   Find out how the following have been heat treated:

● sterilised milk

● UHT or long life milk.
   How long do they keep?
   b   How is yogurt made?

Natural cycles

1 a Copy and complete the passage on the water cycle. Use the words in the box to fill in the blanks.

| evaporated | atmosphere | animals | sea |
| rain | rivers | respiration | |

Water in the Earth's . . . . falls to Earth either as . . . . or snow from clouds. Most of the water arriving on Earth reaches the . . . . as a result of a downhill journey via streams and . . . . . However, some of the water on its way to the sea is taken up by plants, drunk by . . . . or . . . . to the atmosphere. Water may reach the atmosphere from plants and animals by the processes of transpiration and . . . . .

b Draw a neat labelled diagram of the water cycle from the above information.

water in the:	% of total water on Earth
oceans	97.2
glaciers and icecaps	2.15
ground water	0.31
deep ground water	0.31
lakes	0.014
soil moisture	0.005
atmosphere	0.001
rivers	0.0001

table 1 Water on Earth

2 a Look at table 1. Where is the greatest percentage of the Earth's water?
b What percentage of the Earth's water is in the atmosphere?
c Add up the percentages of water shown in table 1. Where is the missing water?

3 Copy and complete the following sentences. Choose your answers from the box after looking at the diagram of the simplified carbon cycle, fig. 2.

Fig. 2

Simplified carbon cycle

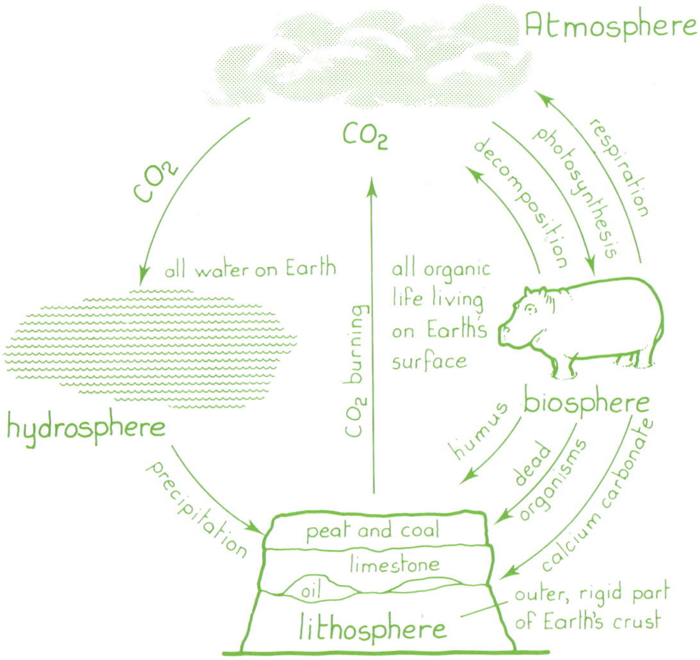

| atmosphere | precipitation | lithosphere |
| photosynthesis | respiration | humus |

a The organic, vegetative content of soil is known as . . . . .
b The . . . . is the outer, rigid part of the Earth's crust.
c Rain is a form of . . . . .
d The . . . . is a gaseous envelope surrounding the Earth.
e Breathing means the same as . . . . .
f . . . . allows plants to trap the Sun's energy, by means of chlorophyll.

4 Write a short description of the life story of a carbon atom, starting with *abc* coal.

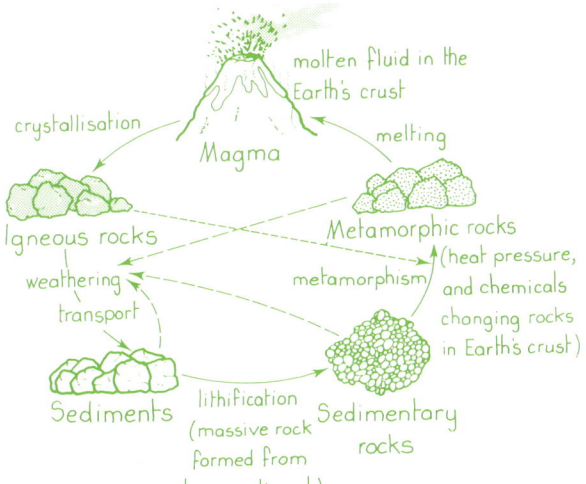

molten fluid in the Earth's crust

crystallisation

Magma

melting

Igneous rocks

Metamorphic rocks

weathering

metamorphism

(heat, pressure, and chemicals changing rocks in Earth's crust)

transport

Sediments

lithification (massive rock formed from loose sediment)

Sedimentary rocks

Fig. 3

The rock cycle

The Earth relies on three sources of energy. These are: *the Sun, heat energy from the Earth's interior* and *the force of gravity*. The development of the Earth has been due to these three energy sources. Study the rock cycle, fig. 3, and answer the following questions.

5 Copy each of these key words from fig. 3 and match it with the correct explanation.

a	Magma	rocks formed by the crystallisation of magma
b	Igneous rocks	rocks changing due to the action of heat, pressure and chemicals
c	Weathering	molten fluid in the Earth's crust
d	Metamorphism	breaking down rocks by the action of wind, rain and other external agencies

6 Write short answers to each of the following questions.
abc
a How can igneous rocks become sediments?
b Explain the term *lithification*.
c What causes sediments to become sedimentary rocks?
d Where does the energy come from to melt metamorphic rocks?
e Can metamorphic and sedimentary rocks become sediments?
f Can igneous rocks become metamorphic rocks?

Heap more trouble

1 Can man disturb the water cycle? Discuss this question in groups of four students. Design a poster to point out of the effect of man on the water cycle.

2 Compost heaps are common in many gardens. They are usually supported by large stones or bricks.
a What are compost heaps?
b What happens in a compost heap?
c What materials can be placed in a compost heap?
d Draw a labelled, cut-away diagram of a compost heap of your own design! Don't forget to include the dimensions or measurements of your compost heap.

River of Death

ASHAMED OF RIVER

Environmentalists say the Rhine must be cleaned up. It has become the longest open sewer in the world. Governments of EEC countries agree to tidy up the river by

RHINE LIVES

The river is much cleaner now, say environmentalists. Fish thrive in the water and river birds have been attracted back to the Rhine. We must continue to safeguard life in the river says

RIVER OF DEATH

Near Basel, in Switzerland, a chemical factory caught fire. During the blaze over 1200 tonnes of chemicals were flushed into the river. The river is now dead to life and may take 30 years to recover.

November 1986

Fig. 4

Cut-outs from The Daily Nag

1 Read the cuttings from *The Daily Nag* shown in fig. 4. They refer to a period of twenty years in the history of the river Rhine.

a What is an open sewer?

b Explain why the river was referred to as the longest open sewer in the world.

c Why was the Rhine called the 'river of death'?

d How might the 1200 tonnes of chemicals have been flushed into the river? A diagram may help your answer.

2 a Fig. 5 shows the passage of the Rhine, through Europe. In which country does the Rhine meet the North Sea?

b Copy fig. 5 and complete the labelling of the countries shown.

c In which countries are the cities shown in fig. 6?

d How many countries were involved in the catastrophe of 1986?

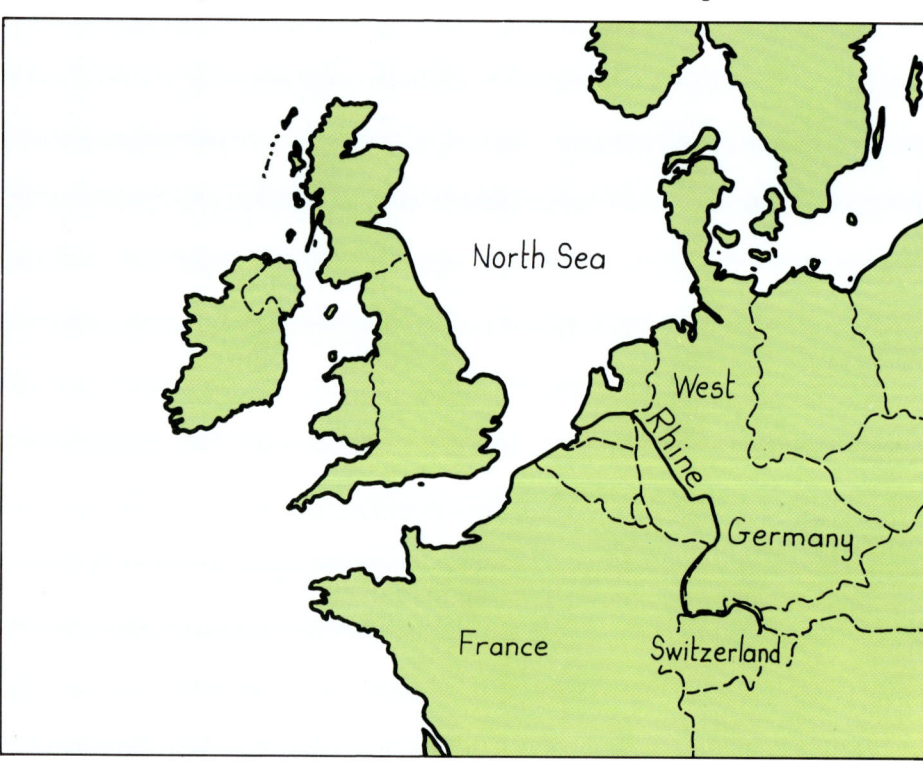

Fig. 5 Europe and the Rhine

3 The chemicals flushed into the Rhine included 12 tonnes of toxic mercury compounds. This mercury passes through the food chain. By the time it reaches animals like otters it has built up into lethal doses. The mercury has become more concentrated.

a Explain what is meant by the term *food chain*.

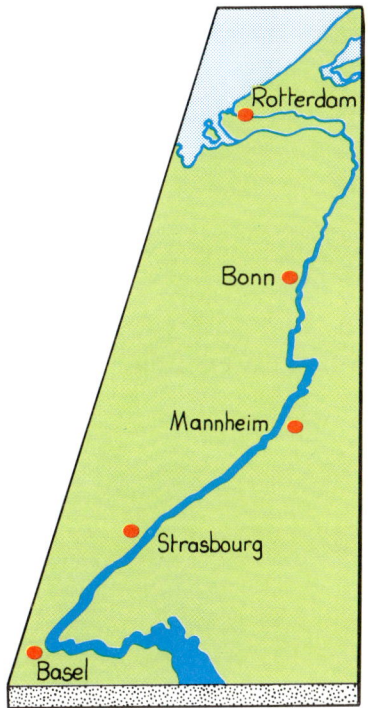

Fig. 6

Cities along the Rhine

b Much of the mercury will hang around in the river silt. What will happen if the silt is disturbed by a passing boat? Do you think it possible to remove all the mercury from the Rhine?

4 In 1976 an accident occurred at a chemical plant in Seveso, Italy. A chemical, dioxin, was released to the environment. As a result many local children suffered severe skin problems and some babies were born badly deformed. Following this accident the EEC introduced special safety rules for making and storing chemicals. These rules apply to all EEC countries.
a In which country did the river of death catastrophe start?
b Is the country a member of the EEC? What effect might this have had on safety standards at the Basel chemical plant?

5 The chemical factory seems to have had only one defence against an environmental accident! This was a small $50\,m^3$ basin in which waste water could be collected. However, at least 500 times this quantity of water was sprayed onto the blaze by firemen! During the blaze drums of chemicals exploded. The chemicals were carried into the Rhine by the water sprayed onto the blaze.
a Estimate the volume of water sprayed by firemen.
b The blaze was fought for about 1000 minutes. Estimate the volume of polluted water entering the Rhine every minute during the catastrophe.
c Basel is approximately 1200 km from the sea and the Rhine flows at a speed of 5 km/h. How long did it take the pollution to reach the sea?
d What extra safety precautions might the chemical firm have taken to prevent such a disaster? Use a diagram to explain your answer.

6 Discuss this question in groups of four students. Imagine you represent the countries involved in the catastrophe. You are meeting exactly one month after the disaster has occurred. Your task is to find ways to repair the damage done to the river. Draw a poster to illustrate your group's ideas. Give a talk to your class describing how you might begin to repair the damage.

In it up to your neck

1 Design a simple a simple experiment to monitor the pollution in the river Rhine. Use a *flow* diagram to show the stages in your experiment! What measurements will you take?

2 Find out about Minamata disease. Minamata is in southern Japan. The disease occurred as a result of a chemical company discharging toxic mercury into the sea.

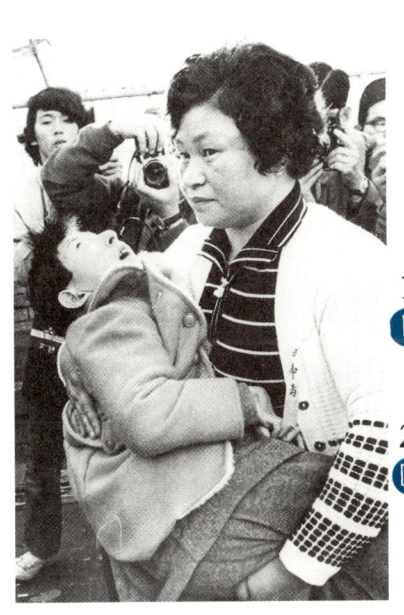

Minamata disease

Weather

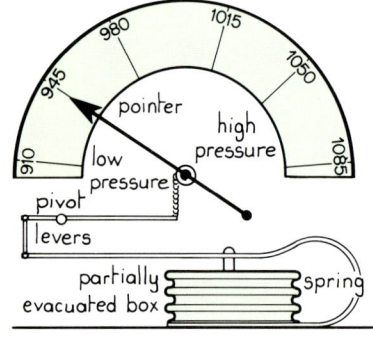

Fig. 7

Aneroid barometer, scale in millibars

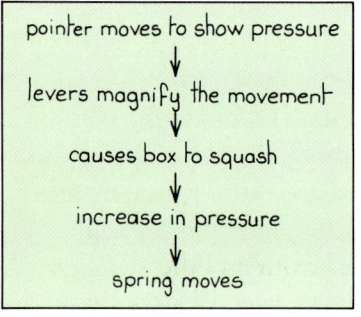

Fig. 8

How does it work?

. . . . the . . . . be fine or . . . . the . . . . be poor, we'll . . . . the . . . . whatever the . . . . together we'll . . . . the . . . . .

1 Copy and complete the passage above. Choose your answers from the box!

weather weather wether

2 a An aneroid barometer measures the atmospheric air pressure. Fig. 7 shows the structure of an aneroid barometer. Fig. 8 is a flow diagram to show how the barometer works. Unfortunately, the stages are jumbled up. Re-draw fig. 8 to show the stages in the correct order.
b Write down the reading of pressure shown, on fig. 7, in millibars.
c Convert the pressure reading to Pascals.
1 millibar = 100 Pascals
d Is this reading a low or high pressure?

3 Rahim has just completed a weather project as part of his science studies. He has taken measurements, over twelve months, of rainfall and temperature at school. His results are shown in table 1.

month	rainfall /mm	mean temperature/°C
January	25	−12
February	20	− 8
March	30	− 6
April	40	2
May	45	10
June	70	15
July	70	20
August	45	20
September	45	14
October	40	4
November	35	− 5
December	25	−12

table 1 Rahim's results

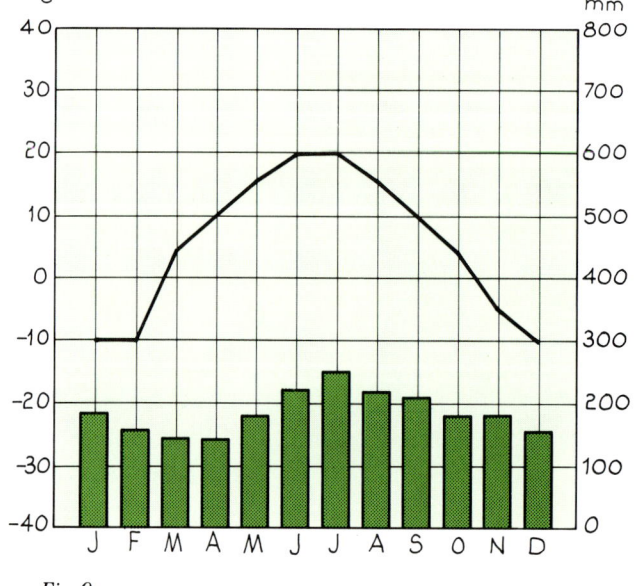

Fig. 9

A climatic graph

a   Use Rahim's results to plot a climatic graph. An example of a climatic graph is shown in fig. 9. The bar chart represents *rainfall*. The graph represents *mean temperature*.

b   Calculate the total annual rainfall at Rahim's school.

c   Calculate the average mean temperature from Rahim's results.

d   Is Rahim's school in the northern or southern hemisphere? Explain your answer.

e   Of what use are Rahim's results?

4   Work in groups of four students to answer this question. *The Daily Nag* is a new, colour, daily national newspaper which will be printed for the first time next month. It has been decided by management that the whole of page 22, in the paper, will be devoted to the weather.

   You have been employed to decide upon the information that should appear on page 22 and also the layout of the weather page. Management have given you the following guidelines to reach a decision:

- the size of the page will be 29 cm × 40 cm

- your layout must appeal to the general public

- the language used on page 22 must be simple

- present as much as is possible of the weather information in easy to understand tables, pictures and maps

- use colour in your layout.

Discuss the task in your group. Decide upon the layout and the information you will include on page 22 of *The Daily Nag*. Write a short summary of your group's ideas. Now produce a full scale mock-up of your weather page.

Clouding the issue

1 a   Write a brief report of how Rahim carried out his weather project. Include descriptions of:

- what Rahim measured

- the instruments he used to take measurements

- how he made sure his results were accurate.

b   What other measures could Rahim have taken which might have improved his project?

2   Collect and write down some more weather rhymes. Do the rhymes ring true? Are there scientific principles behind them? Draw a series of cartoons to illustrate one of your weather rhymes.

Body in the soil

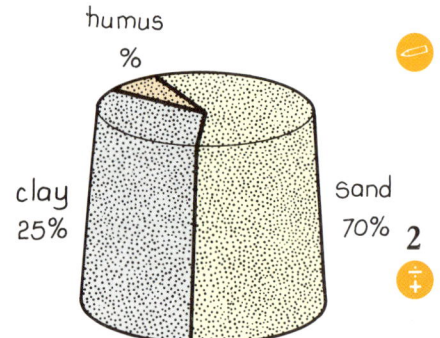
Soil A

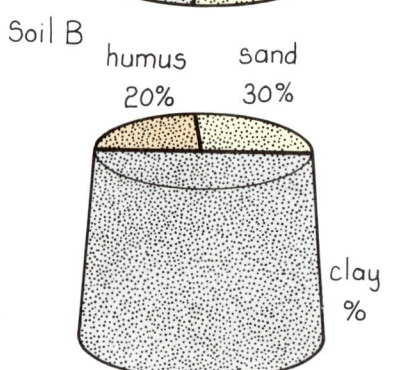
Soil B

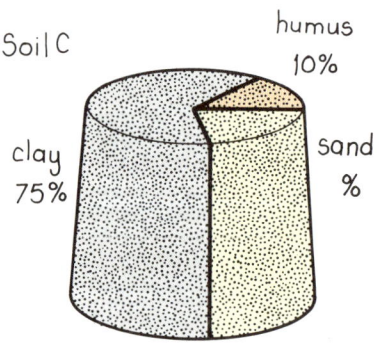
Soil C

Fig. 10

Mud pies

soil	% clay	% humus	% sand
A			
B			
C			

table 1 Soil content

1 Three soil samples are compared in fig. 10.

a What is humus?

b Copy and complete the labelling of fig. 10.

c Which soil has the largest sand content?

d Which sample contains the most clay? Is a high clay content good for soil drainage? Explain your answer.

2 a Copy and complete table 1. Use the information given in fig. 10.

b Plot a bar chart to compare the humus content of the three soils.

c Why do gardeners sometimes add manure to soil?

3 Two Danish peat cutters were working a bog at Tollund Fen on 8 May, 1950. They were surprised to dig up a well preserved body. The body became known as the Tollund man. Murder was suspected by the cutters. Scientific investigation suggested the murder had taken place 2000 years previously! It seems Tollund man may have been sacrificed to the goddess Mother Earth. He was then probably buried in a sacred bog. Sacrifices were meant to bring fertility and good fortune to the local community. They certainly didn't bring good luck to the Tollund man. His body had been immersed in bog water saturated with soil acids. This had helped preserve the Iron Age murder victim. His head is now on display at a Danish museum!

a When was the body discovered?

b Look at the photo of Tollund man. Why was murder suspected? Suggest how the victim might have died.

Surprise in the soil – Tollund man

Grauballe man – another well-preserved Iron Age man, found in Denmark

c What items of clothing appear on the body?
d Which parts of the Tollund man appear well preserved?
e What had caused his body to be so well preserved?
f Why might the victim have been sacrificed?

4 Answer this question in groups of two students. Imagine a bog body has just been discovered in your school grounds. Discuss and design simple experiments to:

● test the acidity of the bog water

● find the water content of the bog around the body.
a Describe briefly what you would do in each experiment.
b Draw labelled diagrams showing the equipment you might use.
c Suggest how you might find out what the victim ate at his last meal.

Soil croppers

1 Aerial photographs give clues as to how land was used in the past. The photo shows the remains of a structure. It was built by the Romans on Hadrian's Wall.
a Suggest what the structure might have been.
b Sketch a plan view of the site.
c Find out about Hadrian's Wall. Why was it built?

Roman remains

2 Find out and write about crop rotation.

3 Many complaints have been made about farmers burning waste cereal straw. Why do farmers burn straw? What hazards arise from straw burning? How could the hazards be minimised? What uses might there be for waste straw?

4 Write a short article to compare natural and artificial fertilisers.

Sounds medical

source	noise level/dB
whisper	25
light traffic	45
conversation	50
radio	60
heavy traffic	80
noisy factory	90
pneumatic drill	95
large thunderclap	110

table 1 Stop that racket!

example	change in noise level/dB
	−5
	−10
serious hearing loss	
double glazing	
car silencer	

table 2 Cutting down noise

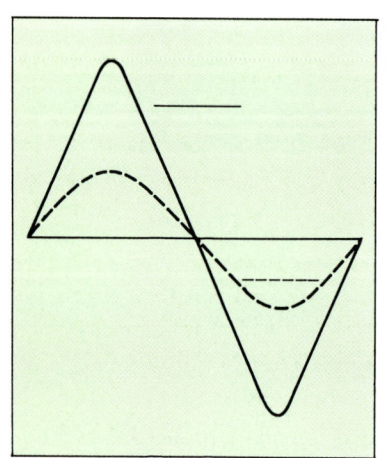

Fig. 12

Comparing loudness

1 Noise seems unpleasant to your ears. Loud noise can be bad for your health. It may cause stress and even hearing loss. Table 1 gives information about noise levels from different sources. Noise levels are measured in decibels, dB.
   a   How much noise does heavy traffic produce?
   b   How many decibels louder than a whisper is a pneumatic drill?
   c   What advice might be given to workers in a noisy factory to protect their hearing?
   d   Draw a bar chart to display the information given in table 1.

2 Look at fig. 11 which gives information about changes in noise levels. For example, it shows that halving a noise level reduces it by 10 decibels.
   a   What happens to the number of decibels when a noise level is doubled?
   b   Write a short passage to summarise the information in fig. 11.
   c   Copy and complete table 2.

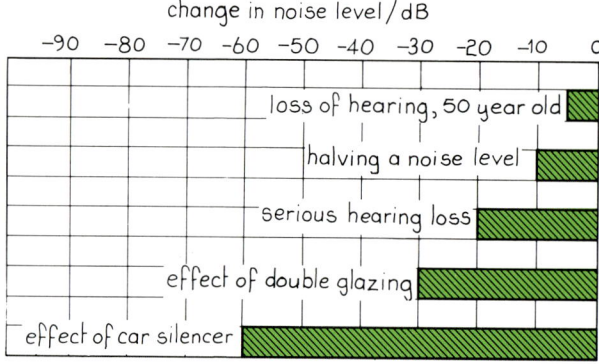

Fig. 11   Changing noise levels

3 It has been suggested that the maximum safe noise level is 85 decibels. Humans experience pain at 120 decibels or above! Fig. 12 is an oscilloscope trace comparing soft and loud sounds. Copy and label the waves in fig. 12 using the words *soft* and *loud*.

4 Hospital noise can be a problem for patients and staff. Fig. 13 compares noise levels in three wards at a British hospital.
   a   What can you say about night-time levels of noise in the wards?
   b   Write down the average noise level in each of the three wards.
   c   What can you say about daytime levels of noise in the three wards?
   d   It is recommended, in the United States, that hospital ward noise levels should be below 35 decibels at night and 45 during the day. Why are these suggested levels so low?
   e   Suggest six sources of noise in a hospital ward.

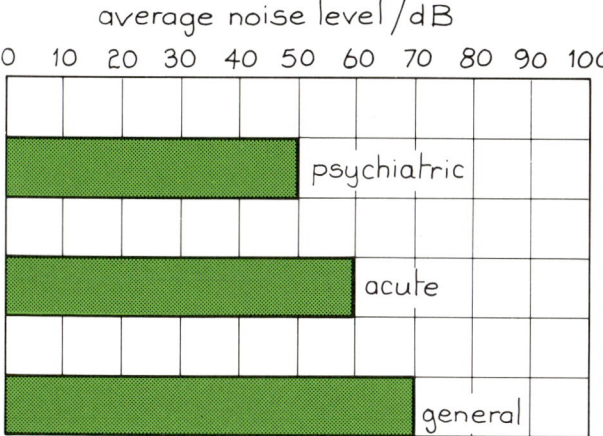

average noise level /dB

night-time noise level /dB

Fig. 13

Warding off noise

5 Answer this question in groups of four students. Imagine you are a hospital medical team. You have received many complaints from patients about noise! Discuss simple ways to cut down the noise levels in the wards.

a Write a set of guidelines for staff giving advice on how to make the wards quieter.

b Design a poster, for display at the hospital, illustrating how noise levels can be kept down.

Sounds awful

1 Humans are said to hear frequencies in the range 20–20 000 Hz. The upper figure is called the **higher threshold of hearing**. This threshold can vary considerably. Design an experiment to find the average higher threshold for your class. You may use any equipment you need. Write a brief report of your design to include:

- a flow diagram showing the stages in your experiment

- a labelled diagram of your experiment in action

- a list of safety precautions you intend to take.

2 a Two girls did a simple experiment to find the speed of sound in air. One clapped her hands 40 m in front of a high wall. The other found the return of the echo to be 0.25 s later. Calculate the speed of sound in air. How might they have made their experiment more accurate?

b Light from a flash of lightning appears instantly. The thunderclap reaches you more slowly, travelling at the speed of sound. How far away is a storm if you hear the thunder ten seconds after seeing the lightning?

3 a Why should the same note played by a piano and guitar sound different?

b Can you explain why a tuning fork sounds louder when its stem is held on the table top?

DDT

Barbados is a small, heavily populated island in the West Indies. It gets its name from the Spanish word for bearded! On Barbados grows the bearded fig tree. The Bermuda petrel, a marine bird, nests on Barbados, getting its food from the sea.

Morocco is a country in North Africa which is over 1000 times larger than Barbados. Several thousand miles of ocean separate the two, fig. 14, Morocco produces large amounts of vegetables and citrus fruits, some of which it exports. However, much damage is caused by locusts chewing and biting the crops. To cut down the number of locusts the Moroccans sprayed their crops with DDT, an insecticide. The DDT gets into an insect either through its skin or by being eaten. Once inside a locust the DDT disrupts its nerve impulses. Eventually, after a bad case of the shakes, the insect dies.

Unfortunately, the Moroccan DDT was carried across the Atlantic Ocean, fig. 15. When it rained the insecticide got into the sea. It then passed along the food chain from plankton to fish to the Bermuda petrel. As it passed along the food chain the concentration of DDT built up. This caused many petrels to die. The DDT built up in the fatty parts of the bird's body. As the bird's fat was used up the insecticide got into the bloodstream, which led to its death.

DDT is a long-lasting chemical. Even when it breaks down it forms another long-lasting insecticide. DDT has been replaced by other insecticides which break down more easily.

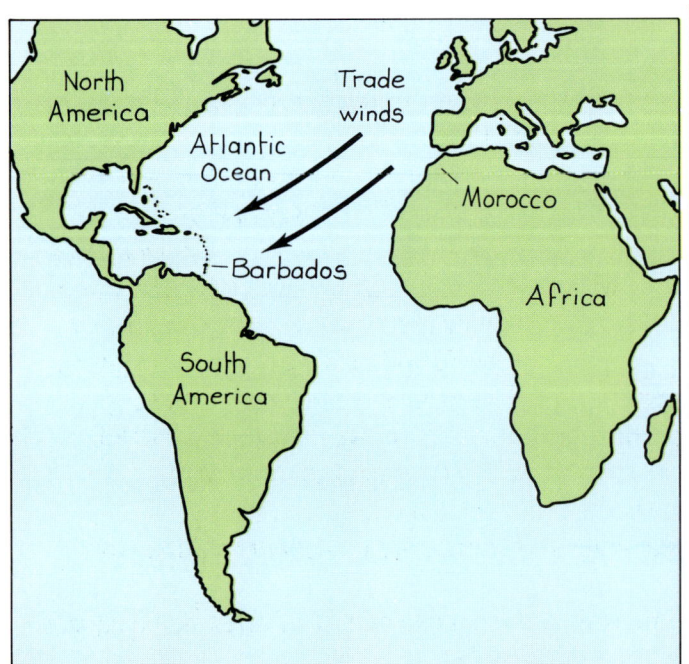

Fig. 14   *Journey to Barbados*

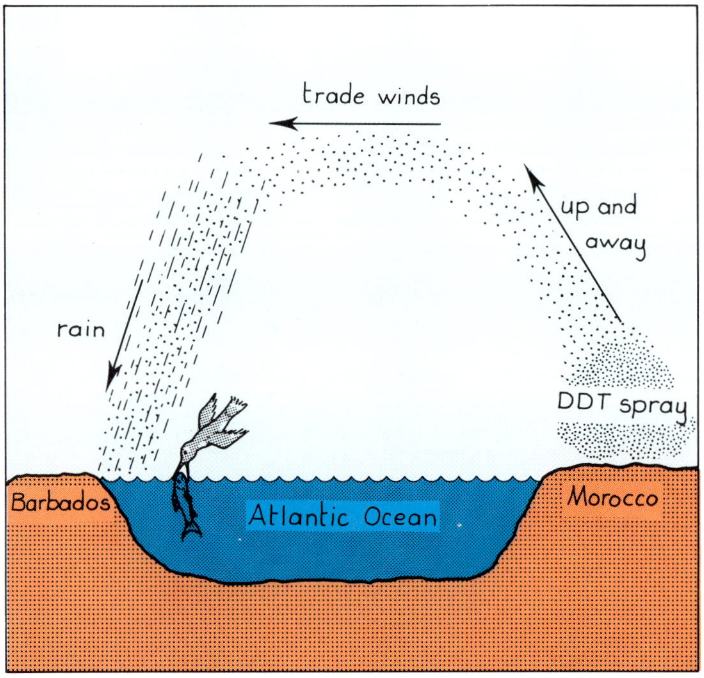

Fig. 15   *Across the Atlantic*

1 a From what does Barbados get its name?
   b Explain why the Moroccans sprayed their crops with DDT.
   c How did the DDT get to Barbados?
   d Explain how insecticides kill insects.
   e Suggest reasons why DDT is no longer used.

2 a Draw a sketch to show how the DDT might have been sprayed onto Moroccan crops.
   b Why don't farmers harvest their crops immediately after spraying?

3 a Draw a flow diagram showing how the DDT reached the petrels.
   b Suggest how the insecticide might have killed the Bermuda petrels.

4 Answer this question in groups of four students. DDT has saved many human lives by killing mosquitos which carry malaria. Insect pests may also be killed by using other killer insects. Greenflies damage crops. Farmers may introduce ladybirds into the greenfly area. The ladybirds produce larvae which feast on the greenflies! Another way of controlling pests is to sterilise male insects which are then released. The males mate with females and pass on their infertile sperm. Discuss:

● the advantages and disadvantages of each method of controlling insects

● the factors which might decide the way insects are to be controlled. Write a brief summary of your group's ideas. Give a talk to your class on controlling pests!

Creepy crawlers

1 Find out about Barbados and Morocco. Write a short article comparing their people and industry.

2 Imagine you live in Barbados. Write a letter to the Moroccan government complaining about high levels of DDT.

3 Clear Lake, in California, is a holiday resort. There were many complaints about the large number of non-biting midges in the area. As a result the lake was sprayed with DDD, an insecticide, in 1949, 1954 and 1957. Nearly all the midges died after spraying. However, their numbers grew quickly in the following years. Over 1000 birds, a colony of western grebes, disappeared from the lake. The grebes fed on fish. Table 1 gives information about the grebe's food chain. It shows that there is 80 000 times more insecticide in the fat of the grebes than in the water.
   a Draw a diagram to show the grebe's food chain.
   b Copy and complete table 1. The insecticide concentration is in parts per million, ppm.
   c Insects may become resistant to an insecticide after a time. Suggest how farmers might overcome this problem.

Insecticide in the:	concen-tration factor	insecticide concen-tration/ ppm
water	1	0.02
plankton	256	
fat of fish	500	
fat of grebes	80 000	

table 1 Poisoning the grebes

Rain, rain go away

1 Anita and Farheen are pen pals. Read their letters about acid rain, figs. 16 and 17. Fig. 18 shows smoking Europe.
  a List four countries that send acid rain to Norway.
  b What fossil fuels are used in power stations?
❓ **c** How is acid rain produced?
  d Draw a flow diagram to show how acid rain might reach Norway from the United Kingdom.

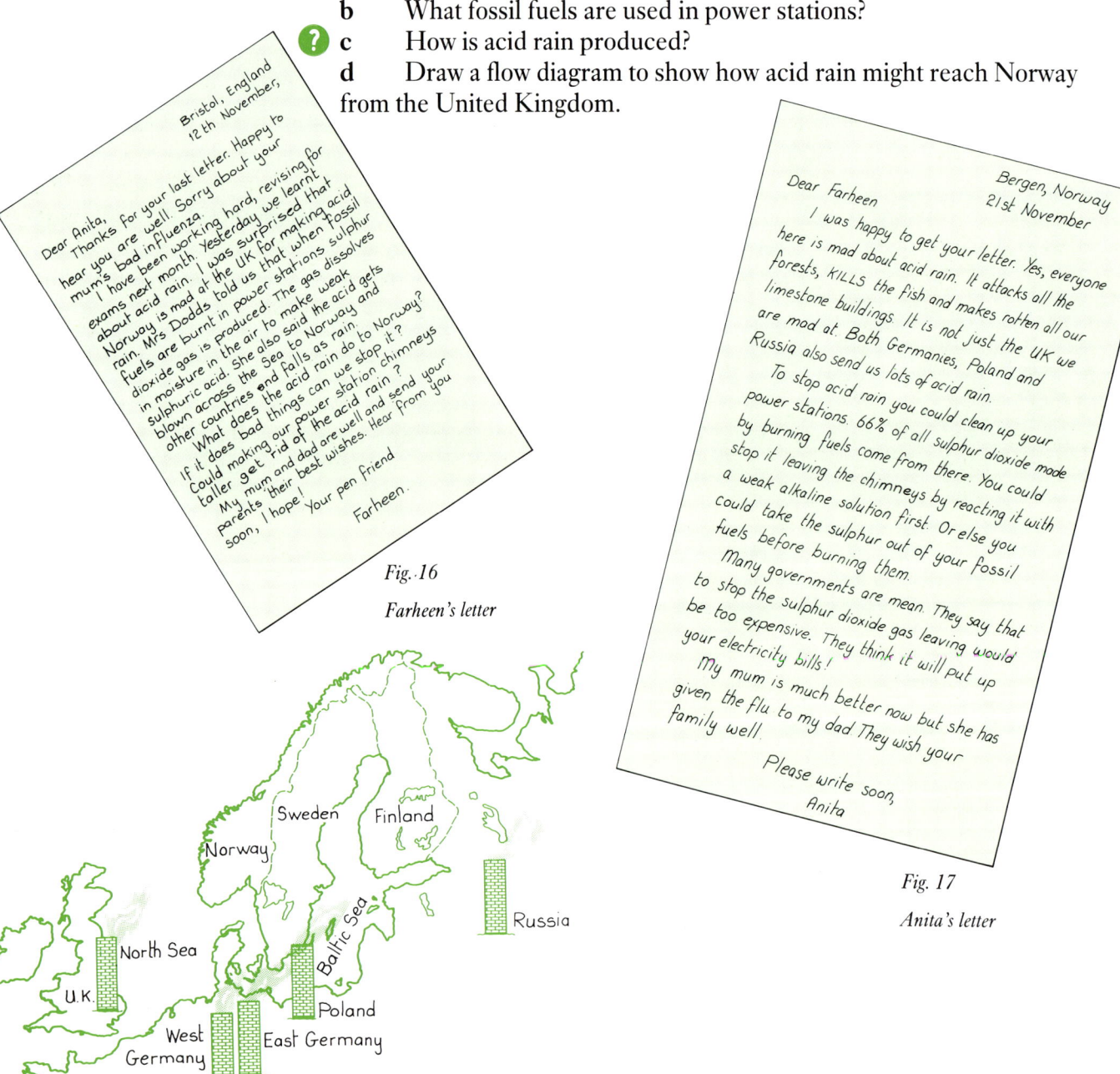

Bristol, England
12th November,

Dear Anita,
  Thanks for your last letter. Happy to hear you are well. Sorry about your mum's bad influenza.
  I have been working hard, revising for exams next month. Yesterday we learnt about acid rain. I was surprised that Norway is mad at the UK for making acid rain. Mrs Dodds told us that when fossil fuels are burnt in power stations sulphur dioxide gas is produced. The gas dissolves in moisture in the air to make weak sulphuric acid. She also said the acid gets blown across the sea to Norway and other countries and falls as rain.
  What does the acid rain do to Norway? If it does bad things can we stop it? Could making our power station chimneys taller get rid of the acid rain?
  My mum and dad are well and send your parents their best wishes. Hear from you soon, I hope!
          Your pen friend
                  Farheen.

Fig. 16

Farheen's letter

Bergen, Norway
21st November

Dear Farheen
  I was happy to get your letter. Yes, everyone here is mad about acid rain. It attacks all the forests, KILLS the fish and makes rotten all our limestone buildings. It is not just the UK we are mad at. Both Germanies, Poland and Russia also send us lots of acid rain.
  To stop acid rain you could clean up your power stations. 66% of all sulphur dioxide made by burning fuels come from there. You could stop it leaving the chimneys by reacting it with a weak alkaline solution first. Or else you could take the sulphur out of your fossil fuels before burning them.
  Many governments are mean. They say that to stop the sulphur dioxide gas leaving would be too expensive. They think it will put up your electricity bills!
  My mum is much better now but she has given the flu to my dad. They wish your family well.

          Please write soon,
                  Anita

Fig. 17

Anita's letter

Sweden Finland

Norway

North Sea

U.K.

Baltic Sea

Russia

Poland

West Germany East Germany

Fig. 18   Smoking Europe

Kinds of lichen	distance from city/ km
5	0
5	5
20	10
35	15
45	20

table 1 Looking for lichens

e Why is Norway mad about acid rain?

f How could acid rain be reduced, according to Anita?

g Anita suggests that cutting down the sulphur dioxide leaving power stations could put up electricity bills. Explain why this might happen.

h Farheen thinks that making power station chimneys taller might get rid of acid rain. Do you think this would work? Use a diagram to explain your answer.

2 About 20 kg per person of sulphur is used every year in the United Kingdom. It is used mostly to make sulphuric acid. Estimate how much sulphur is used every year in the United Kingdom. Assume a population of 50 million people.

3 Lichens are very special plants. They do not have roots and grow well in moderate climates. You can find them growing on walls, trees and rocks. There are over 1000 kinds of lichen growing in the British Isles. They are made of simple green plant cells mixed with threads of fungus. The green plant cells are given shelter by the fungus and the fungus happily accepts the food made by the plant cells! Look at table 1. It gives the results of an investigation, by college students, to find lichens.

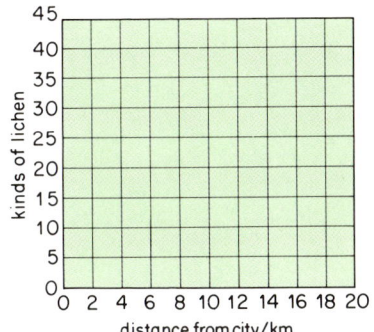

Fig. 19

Finding lichens

a Plot a graph of *kinds of lichen* against *distance from city*. Copy the axes in fig. 19 to start.

b Where did the students find the greatest number of lichens?

c Where did the students find the least kinds of lichen?

d City centres usually have lots of sulphur dioxide pollution. Do lichens like sulphur dioxide?

4 Answer this question in groups of four students. Discuss the point of view shown in the box.

> Who cares about the acid rain in Norway? It's none of our country's business. Let them sort out their own problems. I didn't get where I am today by worrying about other people's problems!

Write a short summary of your group's ideas.

Acid humour

1 Write a humorous letter or poem to *The Daily Nag* newspaper either for or against reducing the amount of sulphur dioxide leaving power stations.

2 Design an experiment to find out about sulphur dioxide pollution in your area. Describe briefly how you would carry out your investigation. What measurements will you take?

It's a gas

1 Look at the British Gas leaflet shown. It tells you what to do if you suspect a gas leak.

a Why should doors and windows be opened?

b Suggest why you should not turn on an electrical switch.

c Where might you turn off the gas supply?

d What should you do if you suspect a gas leak at school?

e Design a poster to tell people what to do when a gas leak is suspected.

HELP
YOURSELF
TO GAS
SAFETY

GAS SAFETY ADVICE

FROM THE GAS PEOPLE

If you suspect a gas leak, here's what to do

1 Put out cigarettes. Do not use matches or naked flames.

2 Do not operate electrical switches (including doorbells) either on or off.

3 Open doors and windows, to get rid of the gas – and keep them open until the leak has been stopped.

4 Check to see if a tap has been left on accidentally, or if a pilot light has gone out.

5 If not, there is probably a gas leak. So turn off the whole supply at the meter and call gas service. The telephone number to call is under 'GAS' in the directory. (Make sure someone is there when we arrive.)

6 If you can't turn off the supply, or the smell continues after you have, or if you have no gas supply, you *must* call gas service, or ask someone else to help do so.

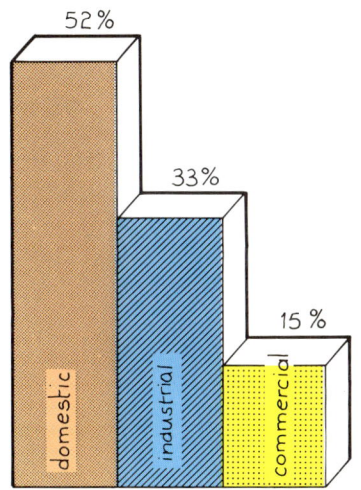

Fig. 20

Using gas in the UK

2 Details of how gas in the UK is used are given in fig. 20.
  a Which is the greatest user of gas?
  b Explain the term *commercial*.
  c Sugest three industrial uses of gas.
  d What might commercial consumers use gas for?
  e Draw a pie chart of the information given in fig. 20.1 per cent should be represented by an angle of 3.6° on your pie chart.

3 Fig. 21 gives information about world reserves of natural gas.
  a Which country has the greatest gas reserves?
  b Calculate the percentage of world gas reserves in Africa.
  c The Middle East has huge reserves of natural gas. However, householders there use bottled gas! Suggest reasons why there is no gas distribution grid in the Middle East.

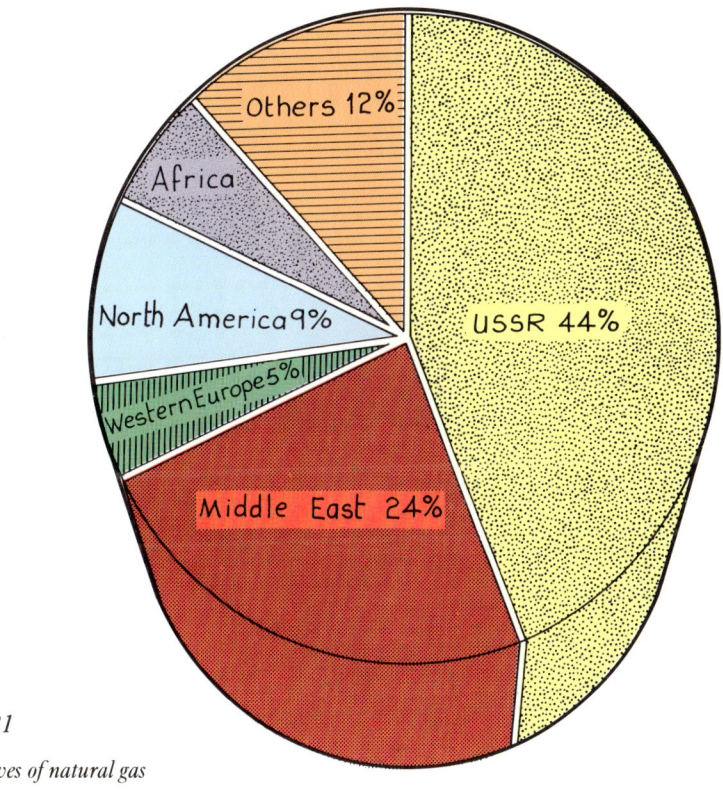

Fig. 21

Reserves of natural gas

More flaming questions

1 Imagine you have just received your gas bill. You believe you have been overcharged. Write a short letter to British Gas to query the bill.

2 a Find out how natural gas is produced and sent to your home.
  b How is coke made from coal? What other products result from the process?

3 It is believed that natural gas may run out during the next century. Write a short article suggesting alternative supplies of gas.

Rumbling old Earth

1 Study table 1 which gives information about the Earth.

a Which region is closest to the surface?

b Where is the crust thinnest?

c Where is the crust thickest?

d Why does the Earth's crust vary in thickness?

e Which region, inside the Earth, is the thickest?

Mount St. Helens 1980

region of Earth	nature and position	thickness of region
crust	thickest under mountains thinnest under oceans solid up to and including surface rocks	maximum 70 km
mantle	hot, molten rock liquid below crust and above outer core	2800 km
outer core	probably liquid below mantle and above inner core	2100 km
inner core	probably solid between outer core and centre of Earth	1400 km

table 1 Structure of the Earth

2 **a** Add the thicknesses of the four regions in table 1. Your answer tells you the radius of the Earth.

b Construct a scale diagram to show the structure of the Earth. Use a circle to represent the Earth. Use smaller circles to represent each region. Write down the scale you have used.

element	% weight
oxygen	47.2
silicon	28.2
aluminium	8.2
iron	5.1
calcium	3.7
sodium	2.9
potassium	2.6
magnesium	2.1

table 2 Elements in the Earth's crust

3 Table 2 gives information about the major elements in the Earth's crust.

a Which element is the second most plentiful?

b What is this element used for in modern technology?

c Draw a bar chart of the information in table 2.

d Does the oxygen in the crust exist as a gas? Explain your answer.

e Which elements in table 2 are metals?

4 Scientists think the Earth's crust is split into several rigid plates. These plates float on hot fluid and move. It is believed the boundaries between

the plates are unstable. Earthquakes and volcanoes seem to occur along these boundaries. Most plate movements happen over a long time. Unfortunately, some happen suddenly. Sudden plate movement can lead to an earthquake at the surface of the Earth.

Look at the journalist's unfinished notes shown in fig. 22. The notes describe the horrific Japanese earthquake of 1923.

a Where was the centre of the disturbance?

b What can you say about the thickness of the crust at the centre of the disturbance?

c What caused Tokyo to lie in ashes?

d What does this tell you about the materials used in the buildings of Tokyo?

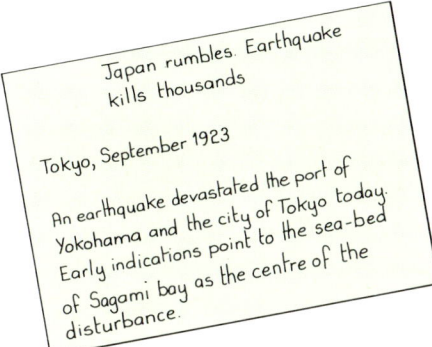

Fig. 22

Japanese earthquake notes, 1923

5 Answer this question in groups of four students. Tokyo lies devastated after the 1923 earthquake. Imagine you are a team appointed by the Japanese government to plan the rebuilding of Tokyo. Your task is to decide upon a set of simple guidelines to give to architects and builders. Your guidelines will make the buildings as safe as possible. You might wish to consider the following points:

● the height of the new buildings

● the materials to be used

● the closeness of the houses to each other

● the position of gas, water and electricity supplies.

Mind rumblers

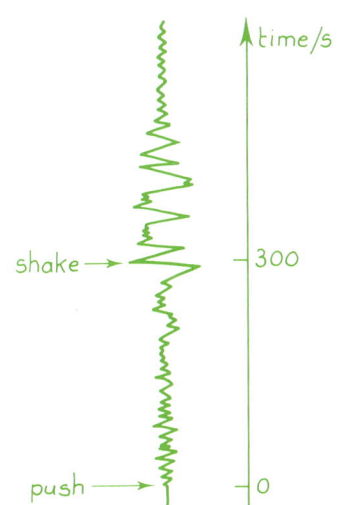

Fig. 23

Seismograph trace

1 Shock waves made during an earthquake can be detected by a seismograph. Fig. 23 is a seismograph trace showing push and shake waves. Push waves are fast, longitudinal waves that travel at 7500 m/s. Shake waves are slower, transverse waves that travel at 4000 m/s.

a Which waves arrive first at the earthquake site?

b Draw two labelled diagrams to show how you can make push and shake waves travel along a rope or slinky coil.

Claws

1 Look at the articles shown in fig. 24.

   a What is iridium?

❓ **b** Why did scientists suggest that an asteroid crashed into Earth?

   c How could a dust cloud have killed the dinosaurs?

   d Why was the asteroid theory 'killed'?

   e Did the dinosaurs all die out at the same time? Explain your answer.

   f Theories change in science. Why should this be the case?

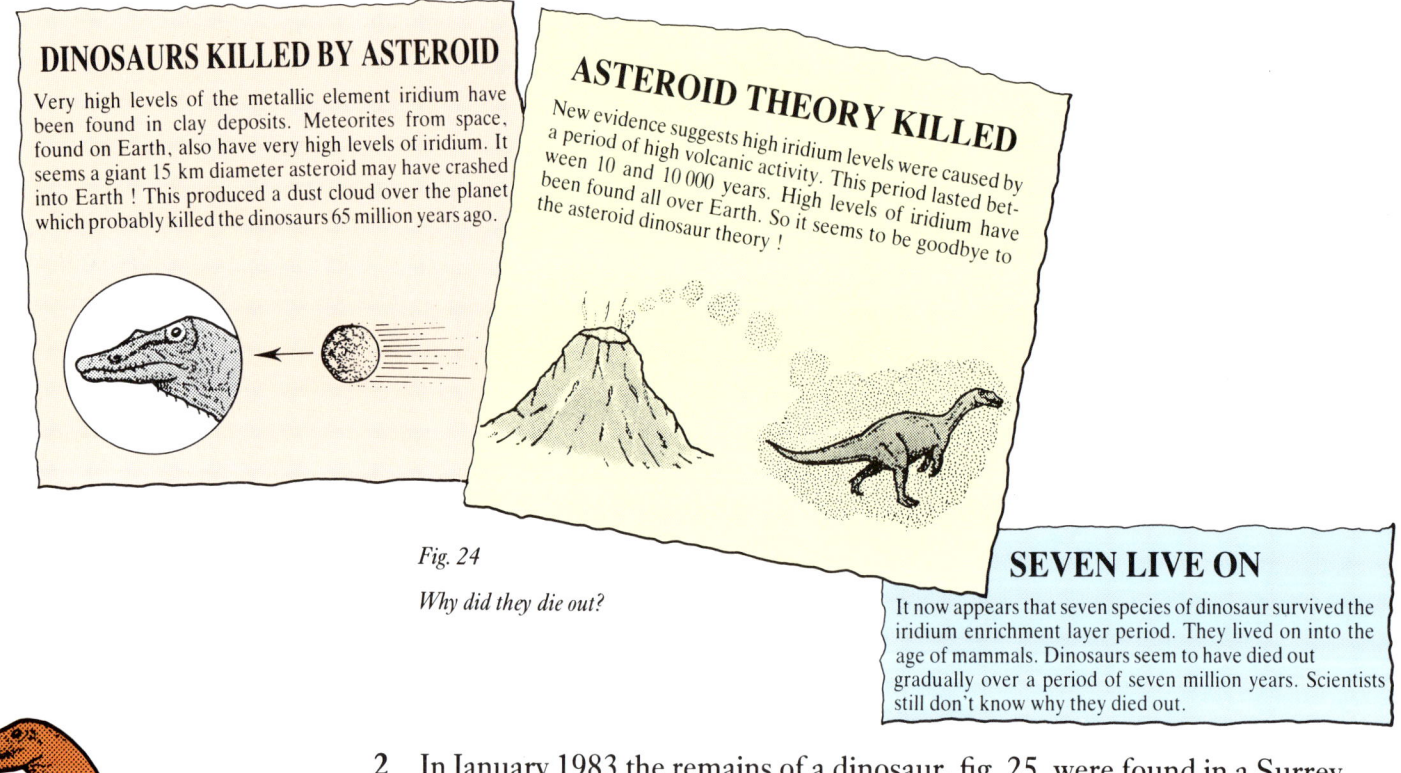

DINOSAURS KILLED BY ASTEROID

Very high levels of the metallic element iridium have been found in clay deposits. Meteorites from space, found on Earth, also have very high levels of iridium. It seems a giant 15 km diameter asteroid may have crashed into Earth ! This produced a dust cloud over the planet which probably killed the dinosaurs 65 million years ago.

ASTEROID THEORY KILLED

New evidence suggests high iridium levels were caused by a period of high volcanic activity. This period lasted between 10 and 10 000 years. High levels of iridium have been found all over Earth. So it seems to be goodbye to the asteroid dinosaur theory !

Fig. 24

Why did they die out?

SEVEN LIVE ON

It now appears that seven species of dinosaur survived the iridium enrichment layer period. They lived on into the age of mammals. Dinosaurs seem to have died out gradually over a period of seven million years. Scientists still don't know why they died out.

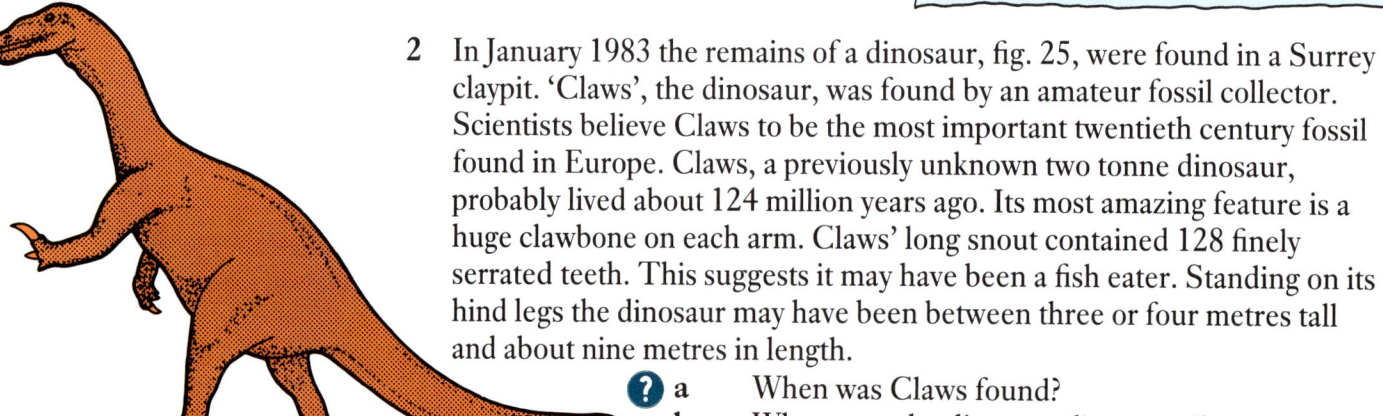

2 In January 1983 the remains of a dinosaur, fig. 25, were found in a Surrey claypit. 'Claws', the dinosaur, was found by an amateur fossil collector. Scientists believe Claws to be the most important twentieth century fossil found in Europe. Claws, a previously unknown two tonne dinosaur, probably lived about 124 million years ago. Its most amazing feature is a huge clawbone on each arm. Claws' long snout contained 128 finely serrated teeth. This suggests it may have been a fish eater. Standing on its hind legs the dinosaur may have been between three or four metres tall and about nine metres in length.

❓ **a** When was Claws found?

   b Where was the dinosaur discovered?

   c In what form were the remains?

   d How many teeth did Claws have?

Fig. 25 Baronyx Walkeri

e   Why might Claws have been a fish eater even though it was a land animal?

f   Why should scientists think Claws is an important fossil find?

3   The first dinosaurs lived about 205 million years ago. They probably looked like small crocodiles. Many different types of dinosaur evolved from these small creatures. Most dinosaurs were either bird-hipped or lizard-hipped. *Diplodocus* was a huge, 10 tonne lizard-hipped dinosaur which was over 27 metres in length. It lived in a herd for protection and probably ate over half a tonne of vegetation daily! It had no teeth or claws for fighting. Its long neck allowed it to graze in thick forests. *Diplodocus* moved about on all fours but could rear up on its hind legs to reach the juicy tree tops. Its brain was surprisingly tiny for such a large animal, being the size of a walnut! Scientists think *Diplodocus* lived about 150 million years ago.

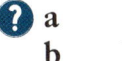

a   Look at fig. 24. How long ago did the dinosaurs die out?
b   When did the first dinosaurs live?
c   For how long did the dinosaurs reign on Earth?
d   When was *Diplodocus* alive?
e   Why did *Diplodocus* live in herds?
f   What evidence is there to suggest that *Diplodocus* was a plant eater?
g   Fig. 26 shows the world 150 million years ago. What do you notice about the world dinosaurs knew? What has happened to the rigid plates that make up the world's land mass since that time?

Fig. 26

Dinosaur's world, 150 million years ago

4   Discuss this question in groups of four students. The climate during the time of the dinosaurs was warmer. Even Siberia had a mild climate then. Plants and animals probably thrived in the warm seas of the time. Generally, the climate was perfect for those great big, hungry dinosaurs. Why then did they die out?

a   Design and draw a poster showing your group's ideas.
b   Write a follow up article to fig. 24, summarising your group's thinking.
c   Draw a follow up cartoon to fig. 26 featuring Claws!

Plod on

1   Find out about three of the ancient animals in the box. Write a short article on each. Illustrate each article with a diagram of the creature.

Archaeopteryx	*Ichthyosaurus*	*Plesiosaurus*
Stegosaurus	*Triceratops*	*Tyrannosaurus*

2   Fossils are organic traces which have been buried by natural processes and preserved permanently. Find out and write about them.

Titan

Saturn, rings true

Titan, the Earth in deep freeze

Saturn is the sixth planet in order of distance from the Sun. Have you seen Saturn? It can be seen with the naked eye. Christian Huygens, in 1659, was the first to notice that Saturn was surrounded by a separate system of rings. Saturn has several moons or satellites, the largest of which is **Titan**. Huygens discovered Titan in 1655. Scientists believe that clues to primordial or early Earth may be found on Titan.

Recent space probes have found a nitrogen atmosphere more massive than Earth's on Titan. It does not seem to have its own magnetic field or magnetosphere. Titan appears to be made up from equal amounts of ice and rock.

As well as nitrogen, methane has been found in Titan's atmosphere. The action of the Sun's light on the methane has led to a range of other chemicals being formed. The nitrogen and methane seem to have reacted to produce hydrogen cyanide. The discovery of hydrogen cyanide has delighted scientists. Hydrogen cyanide is a key substance in the synthesis of amino acids and bases present in nucleic acids. Amino acids and bases are important biological compounds which appear on Earth. It seems that Titan is like the Earth in deep freeze!

Water can exist as a solid, liquid or gas. On Earth it exists in the form of ice, water and water vapour. Methane seems to exist as a solid, liquid and gas on Titan. Titan's methane may play the same role as Earth's water! Could there be rivers, lakes and seas of methane on Titan? There seem to be clues on Titan which may help us to understand the evolution of early life on Earth.

1   Copy each of these key words from the passage and match it with the correct explanation.

   a   Primordial   surrounding magnetic field.
   b   Satellite   development from earlier forms.
   c   Magnetosphere   moon, revolving round a planet.
   d   Synthesis   early, from the beginning.
   e   Evolution   combining of parts, elements or compounds.

2   a   Who discovered Titan and in what year?
   b   How might you identify Saturn through a telescope?
   c   Why is Titan of interest to scientists?

3   The water cycle on Earth is necessary for the wellbeing of life. A similar methane cycle may occur on Titan. Draw a methane cycle for Titan.

4 a Which substance, discovered on Titan, could lead to the synthesis of amino acids?

b Which gas, very evident on Earth, appears to be missing on Titan?

5 Look at table 1. It gives information about eight of Saturn's moons.

a Which satellite is the smallest?

b Which satellite is the largest?

c What do you notice about the densities of the satellites?

satellite	orbital radius/R_s	satellite radius/km	density/kg $\frac{}{m^3}$
Mimas	3	195	1200
Enceladus	4	250	1100
Tethys	5	525	1000
Dione	6	560	1400
Rhea	9	765	1300
Titan	20	2560	1900
Hyperion	25	145	–
Iapetus	60	720	1200

table 1 Saturn's satellites. R_s is the radius of Saturn.

6 Using A4 graph paper, draw a scale plan view of the eight satellites orbiting Saturn. You may assume circular orbits.

7 Do you think life could exist on Titan? Discuss your ideas in groups of four students. Write a brief summary of your group's conclusions.

Titanic struggle

1 Galileo first observed Saturn's rings in 1610. At that time he could not see the rings separate from the planet. Galileo Galileii lived in Italy between 1564 and 1642. He was an astronomer and physicist. Find out and write about his discoveries.

2 a Write down the radius of Titan in km.

b Calculate r, Titan's radius in m.

c Calculate Titan's volume in m^3.

$$volume = \tfrac{4}{3}\pi r^3$$

d Write down the density of Titan in kg/m^3.

e Calculate Titan's mass in kg.

$$mass = density \times volume$$

Lunar-cy

Shine on, silvery Moon

Saturday Night

On Saturday night I lost my wife, And where do you think I found her? Up in the Moon, singing a tune, And all the stars around her.

Fig. 27
Traditional rhyme

Dear Nevil,
On the Earth there is both heat and light. As a result the Earth is inhabited by people. The Moon, our satellite, also has heat and light. Looking at its surface there does seem to be a soil which appears every bit as good as our own, if not better! How can anyone argue with the fact that there must be inhabitants of one type or another, on the Moon?
Yours truly,
Bill Herschel

Fig. 28
Arguing the case

Can there be a more beautiful sight than the Moon shining brightly in the night sky? Nursery rhymes, fig. 27, refer to people living on the Moon. Did people really believe the Moon had inhabitants?

Aristotle lived between 384 and 322 BC. He believed the Moon was perfect, with a smooth surface. He thought it travelled in its own crystalline sphere around the Earth. According to Aristotle the dark patches on the Moon were reflections of Earth's rough surface. His theory was believed for over two thousand years!

Early in the seventeenth century Galileo studied the Moon's surface. He used the newly invented telescope. Galileo found the surface to be imperfect. He saw plains, mountains and even what he thought were seas. Cleverly, he worked out the heights of mountains from the shadows they cast. Many famous scientists of the time believed the Moon to have inhabitants. They thought that with large enough telescopes they might see the inhabitants walking around! Sir William Herschel was a great astronomer who discovered the planet Uranus in 1781. Read his letter to Nevil Maskelyne, the Astronomer Royal of the day, shown in fig. 28.

In 1969 Neil Armstrong was the first man to land on the Moon. The rock samples he brought back showed no trace of life. Scientists now believe the Moon to be lifeless.

1   **a**   What did Aristotle believe the surface of the Moon was like?

?   **b**   Sketch what Galileo might have seen when he looked at the Moon through his telescope.

     c   Are there really seas on the Moon?

     d   Why do scientists now believe the Moon to be lifeless?

2   Study Herschel's letter to the Astronomer Royal, fig. 28.

     a   Herschel believed life existed on the Moon. List the arguments he used.

?   **b**   Nevil Maskelyne, the Astronomer Royal in 1780, disagreed with Herschel about life on the Moon. Imagine you are Nevil Maskelyne. Write a reply to Herschel countering his arguments.

3 Answer this question in groups of four students. Imagine you are a team of scientists. You have been asked to plan a mission to the Moon to build Moonbase I. It is to be *permanent* and *self-sufficient*. What materials would you need? It might help you to think about the following points:

● the Moon has neither water nor an atmosphere

● you weigh less on the Moon

● clothing must allow for hot days and cold nights

● waste must be disposed of

● power and food must be produced.
Draw a poster showing the materials you need to set up Moonbase I.

4 Devise an experiment to carry out at Moonbase I. Choose from the following titles:

● growing plants

● testing clothing for lunar conditions

● dealing with waste products.
Draw a flow diagram to summarise your experiment. What measurements will you take? What equipment will you need?

Lunar boosters

1 a Collect and write down your favourite lunar nursery rhymes.
 b There are many superstitions about the Moon. It has been believed that the behaviour of people and animals is affected by the Moon. Write about one such superstition.

2 Look at table 1. It gives information about four of Jupiter's Moons. They were first discovered by Galileo and are called the Galilean satellites.
 a Plot a graph of *orbital time* against *distance*. Copy the grid in fig. 29 to start with.
 b Amalthea is another Moon of Jupiter. It is 110 000 km from Jupiter. From your graph predict the time it takes to orbit Jupiter.

satellite	distance from Jupiter /km	orbital time /days
Io	350 000	1.8
Europa	600 000	3.6
Ganymede	1 000 000	7.2
Callisto	1 810 000	16.7

table 1 Jupiter's Galilean satellites

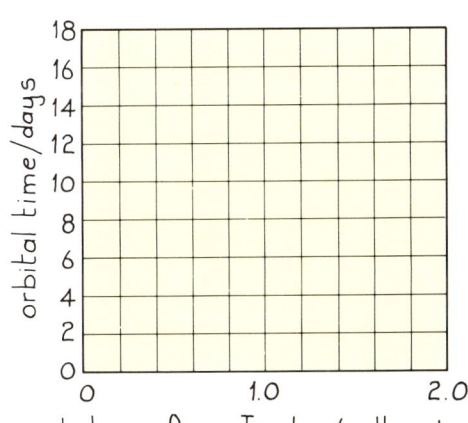

Fig. 29

Circling Jupiter

Rocky 5

Marie, Jamie and Ann carried out an investigation to find out if rocks soaked up water. They collected five dry samples of different rocks from Mr Campbell, their teacher. The students found the mass of each rock using a lever arm balance. Each rock was then soaked in water for 72 hours. They then found the mass of the rocks after soaking. Part of Marie's unfinished report is shown in fig. 30.

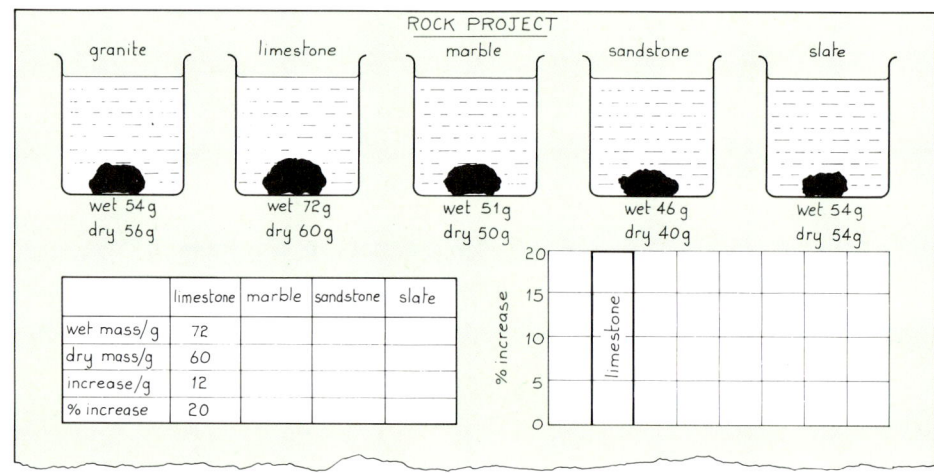

ROCK PROJECT

granite | limestone | marble | sandstone | slate

wet 54 g / dry 56 g — wet 72 g / dry 60 g — wet 51 g / dry 50 g — wet 46 g / dry 40 g — wet 54 g / dry 54 g

	limestone	marble	sandstone	slate
wet mass/g	72			
dry mass/g	60			
increase/g	12			
% increase	20			

Fig. 30 Marie's unfinished report

1 a Which rock soaked up the most water?
  b Why did they begin their investigation with dry rocks?
  c Why did Marie not include granite in her results table?
  d Suggest how the students might have made their wet samples dry again at the end of their investigation.

2 a Copy and complete Marie's results table.
  b Copy and complete her bar chart. Use graph paper.
  c What conclusions might Marie have come to about the investigation?

3 a How might the students have made their results more accurate?
  b Use a diagram to explain where the water went to in the rocks.
  c Suggest practical uses for each type of rock used by the students. Present your answer as a table. Your headings might be *rock* and *use*.

4 Answer this question in groups of two students. How does the mass of a lump of limestone, soaking in water, change over a week? Design an experiment to find out the answer to the question.

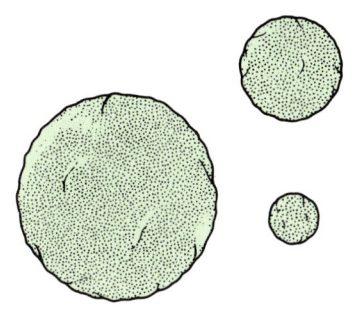

Fig. 31

Jamie's rocks

sample	average diameter/mm
small	
medium	
large	

table 1 Measuring rocks

a   What measurements will you take?
b   Briefly describe what you will do in your investigation.
c   Draw a diagram to show all the equipment you intend to use.

5   Jamie collected three spherical rock samples on a school field trip. They are compared in fig. 31. Measure the diameter of each sample twice. Calculate the average diameter of each rock. Copy and complete table 1. What may have caused the rocks to be spherical? Suggest where Jamie might have found his samples.

Rock around the clock

1   Professor Tardis compared two dry soils. He placed equal amounts of sandy and clay soil in two funnels. Simultaneously he began to pour water into each. His experiment, one minute after starting, is shown in fig. 32.
a   List the equipment he used. It's not all shown in fig. 32!
b   Why did the Professor use cotton wool in his funnels?
c   Which soil is badly drained? Identify the soils in fig. 32.
d   How did the Professor make his experiment fair?

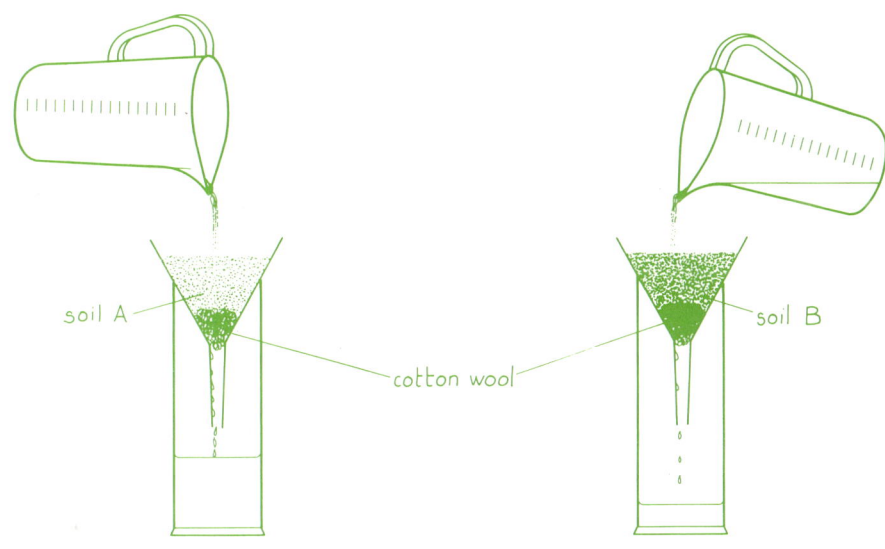

Fig. 32

Comparing soils

2   Marie wanted to find out what happened to water when it froze. She put increasing amounts of water into four small measuring cylinders. They were placed in a chest freezer overnight. Marie noted the volume of ice in each cylinder the next morning. Her results are shown in fig. 33.
a   What happened to the water when it froze?
b   What effect might this process have on rocks which have soaked up water?
c   Explain why Marie used a plastic, not a glass, measuring cylinder.

Fig. 33

Marie's results

Find the joker

```
F X A W L M N G O N I O N
E J B A B E O M A I L O H
R C L D E L P H G J I M K
M F O B M A N L I T C C L
E B O K R T W X A Z E N P
N O D N E T X R Y C R K D
T N M E P E I B L Y S I S
I E T J S P S S V T G D S
N H I H S X L V S E A N U
G K L N M S L T S U Z E W
Y E A S T Y E T O N E Y Z
R R Q P V A C U O L E W A
T R A E H O S M O S I S G
```

Fig. 34

World puzzle

The answers to the following clues are in the word puzzle, fig. 34. Can you find them? Write down your answers. Each clue is in two parts:

- the first part of a clue is written

- the second part is a clue to the first letter of the answer. Fig. 35 shows you the directions in which you might find your answers. One answer is a **joker**. It is not in the word puzzle. Can you find the joker?

- microscopic organism; 1

- red liquid found in veins; 2

- part of a vertebrate skeleton; 2

- all living organisms are made of these; 3

- done to food in the stomach and bowels; 4

- breaking down sugar to alcohol; 6

- organ that pumps blood; 8

- organ that gets rid of urine; 11

- parasites found in hair; 12

- edible bulb that makes you cry; 15

- process of water flowing through a semi-permeable membrane; 15

- a light surgical knife; 19

- fertilisation occurs when one meets an egg; 19

- these help you bite; 20

- joins a muscle to the skeleton; 20

- substance of the body; 20

- process of water evaporating from the above-ground parts of a plant; 20

- at the centre of a cell; 22

- blood flows through these; 22

- fungi used to make wine and beer; 25

Fig. 35

Directions

Mind Sweepers

1 Start with the *marked* square on the supergrid, fig. 36. Answer each clue in turn. Find your answer on the grid. Write it down. The last letter of your answer becomes the first letter of your next answer! You can move *down* or *up*, *forwards* or *backwards*. Don't jump a square.

- upper limb, attached to shoulder
- warm-blooded animal
- sticks out from plant stem, green and flat
- nourishment
- large extinct reptile
- third word in previous clue
- oval, laid by a bird
- contained by chromosomes
- these grow into plants
- found in bread, cereals and potatoes
- grows on your head but not on mine
- the colour of blood
- more of those large, extinct reptiles
- the upper layer of earth

Y	R	M	G	G	R	B	L	I
X	A	A	E	E	N	O	S	O
O	O	M	L	N	I	D	A	S
D	F	MI	E	S	E	U	R
I	A	A	T	P	E	R	I	A
N	E	L	M	E	E	T	A	H
O	S	A	U	R	D	S	R	C

Fig. 36

Supergrid

2 Can you solve Jimmy's riddle? The answer fits into fig. 37.

My first is in lens and also in mirrors,
My second is in seismic but not in mixtures,
My third is in therm and also in shock,
My fourth is in hour and also in clock,
My fifth is in toe but not in time
My sixth is in clay and also in lime,
You're there everyday, I'm sure you have guessed
It must be the place that you love the best!

Fig. 37

Rhyme and reason

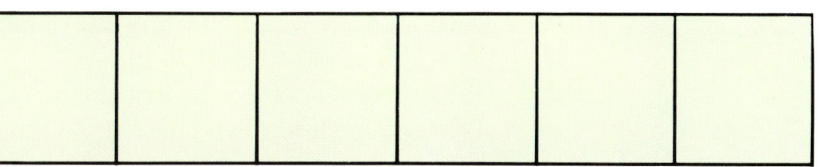

Rubella

Sharika, Jane and Mark did a group project on rubella as part of their science studies. Mark produced a set of cartoons, Sharika wrote an article, fig. 38, on the disease. Jane produced an article on viruses, fig. 39.

3 March

Rubella

Rubella is a mild, infectious disease. It is a virus infection that has a three week incubation period. When you catch rubella the glands behind your ears and those down the side of your neck get swollen. You also get a headache, sore throat and generally feel under the weather. You get a pink rash for two or three days. This rash is really lots of small, raised spots which cover your face and trunk. Sometimes, grown ups with the disease get painful joints.

No treatment is usually necessary.

Rubella is a problem when caught by a pregnant woman in the first three months of her pregnancy. This is because the virus can pass across the placenta to the foetus. This means it can damage the unborn baby. The damage to the baby means it could be born with cataract eye, deafness, defects of the heart or microcephaly. Microcephaly means the baby's head is small, it could be mentally deficient and it could get convulsions.

To stop this horrible damage to the unborn baby all females should be immunised against rubella. This is when you get a simple injection which protects you against the disease. Side effects are rare. At worst you might get mild symptoms of the disease.

Sharika   Form 4L

Fig. 38

Sharika's article

Jane 4L   VIRUSES

Viruses are tiny organisms that are smaller than bacteria. You can't see them under an ordinary microscope. Several million would fit across this page. Most viruses are supposed to have simple shapes. They aren't cells because they don't have a nucleus. On the outside they have a protein wall. On the inside they have a coiled up strand of something called nucleic acid.

Viruses are horrible thieves! This is because in order to reproduce they steal the insides from a living cell. They kill the living cell in the process.

Different types of virus will attack different types of cell. Viruses cause diseases like flu, the cold, rabies, mumps, polio, poliomyelitis, AIDS and rubella. Funnily enough antibiotics don't kill viruses. Immunisation is when a doctor puts a small amount of vaccine into your body. This protects you against the virus. The vaccine contains proteins called antigens. When the antigens are injected in you they make your white blood cells make antibodies. Your antibodies are great because they kill those horrible viruses.

Fig. 39

Jane's article

1 Read Sharika's article shown in fig. 38.

a What is rubella?

b Write down the symptoms of the disease.

c What is the treatment for rubella?

d Why is rubella said to be a problem?

e What effect can rubella have upon the unborn child?

f What is meant by microcephaly?

2 Study Jane's article on viruses, shown in fig. 39.

a What is a virus?

b How big is a virus?

c Describe the structure of a virus.

d How do viruses manage to reproduce?

e Do antibiotics kill viruses?

f Can antibiotics be used against rubella?

3 Both Jane and Sharika mentioned immunisation in their articles.

a Explain what happens when you are immunised.

b How can immunisation against rubella prevent damage to the unborn child?

c Before being immunised by a doctor you might be given a blood test. Why?

4 Answer this question in groups of three students. Imagine you are Sharika, Jane and Mark! Discuss and decide upon a publicity campaign to inform your fellow students about the effects of rubella. Your campaign should involve drawing at least two posters to display around school. You might also give a talk on rubella to your class.

5 a List six occupations where pregnant women come into regular contact with young children.

b Do pregnant women have a higher risk of catching rubella if they are in frequent contact with young children? Explain your answer.

Foxy tales

1 a In 1917 an outbreak of influenza killed more people than did the fighting in World War I. Find out and write about the causes and prevention of influenza. Is it sensible to go to school if you have the flu?

b Nearly 80 per cent of babies were immunised against whooping cough in the mid 1970s. However, only about 50 per cent of babies are now being immunised against the disease. Why?

2 In Switzerland vets have begun a campaign to immunise wildlife against rabies. Foxes are the main carriers of rabies in Europe. The campaign uses chicken-heads that have been dosed with live vaccine as bait for the foxes.

a Why has live vaccine been used in the campaign?

b What special problems could arise from the use of live vaccine?

c The Swiss campaign has concentrated on valley areas. To rid Europe of rabies would be a major scientific achievement. However, much of Europe is lowland. What extra problems might arise if the campaign is extended to lowland areas?

Cousin Lesley's tum

Cousin Lesley took a pill
That made her go invisible.
Perhaps this would have been all right
If everything was out of sight.

But all around her stomach swam
Half-digested bread and jam,
And no matter how she tried,
She couldn't hide what was inside.

In the morning we often noted
How the toast and porridge floated,
And how unappetising in the light
Was the curry from last night.

Some Gruyère had fallen victim
To her strange digestive system,
And there seemed a million ways
To digest old mayonnaise.

We were often fascinated
By the stuff left undigested,
A mish-mash of peas and jelly
Drifted round our cousin's belly.

Certain bits of Cornish pastie
Looked repugnant and quite nasty,
While the strawberries from last year
Were without the cream, I fear.

And at dinner, oh dear me!
What a disgusting sight to see.
Chewed-up fish and cold brown tea
Where Cousin Lesley's tum should be.

Brian Patten

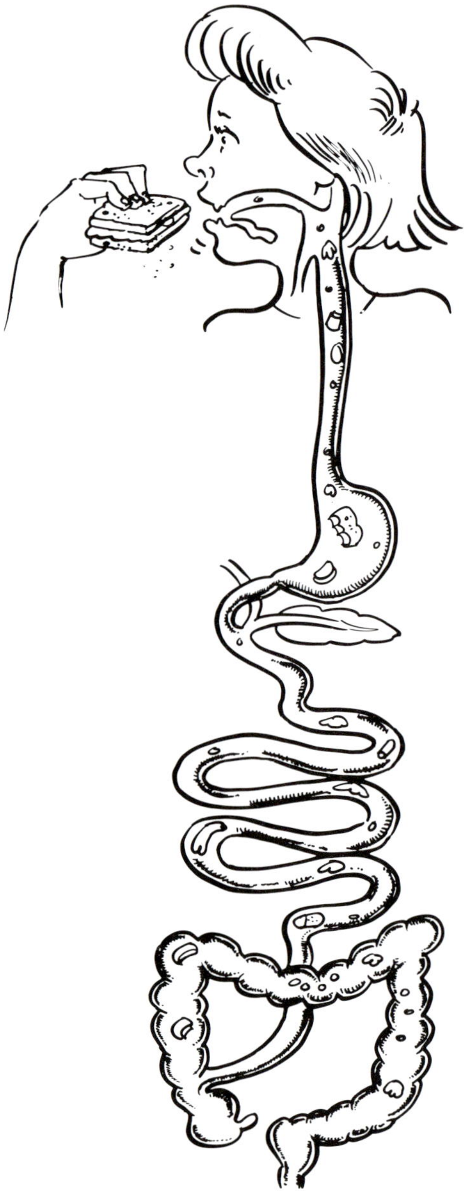

Fig. 40

Food for thought

1 Study fig. 40.
   a   Why did Cousin Lesley become invisible?
   b   Describe what her family saw in the morning.
   c   What is Gruyère?
   d   Describe the delightful sight at dinner.

2   Copy and label the diagram of Cousin Lesley. Use the labels in the box.

anus	colon	gullet	ileum
mouth	pancreas	stomach	salivary gland

age range/years	B_{12} intake/μg
0–1	
1–3	
4–6	
7–10	
11–16	

table 1 Recommended vitamin B_{12} intake

3 Vitamin B_{12} is found in most foods of animal origin. It dissolves in water and is easily absorbed by your body. B_{12} is essential for the production of red blood cells. If you do not get enough of the vitamin you produce fewer red cells. This can lead to the uncommon condition of B_{12} anaemia. Fig. 41 gives details of recommended daily intakes of the vitamin.

a Suggest three sources of vitamin B_{12}.

b Copy and complete table 1. Use the information in fig. 41.

c Suggest why B_{12} anaemia is uncommon.

d Pregnant women and breast feeding mums need 4.0μg of B_{12} per day. Other adult females need only 3.0μg daily. Why?

e B_{12} deficiency occurs in people who have had their stomachs removed. A similar deficiency occurs in those who have had an operation on their terminal ileum. What does this observation suggest? How might such a B_{12} deficiency be overcome?

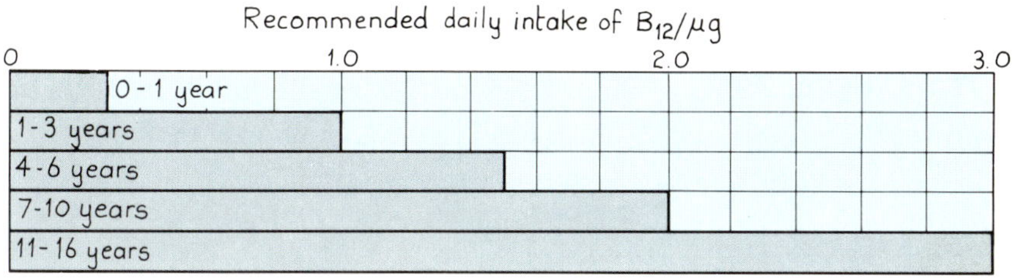

Fig. 41

Vitamin B_{12} intake

Recommended daily intake of B_{12}/μg

0 — 1.0 — 2.0 — 3.0

0-1 year
1-3 years
4-6 years
7-10 years
11-16 years

Readers' digest

1 You have just eaten a tasty piece of pizza. It now begins its long journey along your gut. Write a poem with the title *'Journey To The End'* describing its passage. Illustrate your poem with a diagram.

2 Answer this question in groups of three students. Your salivary gland produces amylase, an enzyme. This begins to break down starch in your mouth. Does saliva produced before eating break down starch more slowly than that produced during eating? Discuss and design an experiment to find out. Don't forget your mouth starts watering before you start to eat! You might consider using:

● a starch solution

● iodine solution to test for starch.

a List the equipment you would use.

b What measurements would you take?

c Describe what you would do.

d What do you expect to find out?

3 Suggest explanations for the following statements:

● faeces smell bad

● fibre in the diet is good for you

● after going to the toilet you should wash your hands.

Peace pipe

American Indians smoked tobacco in the pipe of peace. They also chewed tobacco and used it as snuff. Francisco Fernando, a Spanish doctor, brought the plant to Europe in 1558. It was thought to be a cure for gout, indigestion and scurvy. In 1586 Ralph Lane, the first governor of Virginia, gave Sir Walter Raleigh a smoking kit for a present. It is said that one of Raleigh's servants was horrified to see smoke surrounding Sir Walter. The servant doused his master with the contents of a tankard, to put out the flames! Sir Walter certainly has a lot to answer for as he helped to make smoking popular.

Early seventeenth century smokers did not have it so easy. King James I, of England, hated the habit. In 1604 he published his 'Counterblaste to Tobacco'. Pope Urban VIII threatened to excommunicate anybody bringing tobacco into a church. In France, Louis XIII made it illegal to sell tobacco except on doctor's orders.

Fashions in using tobacco have changed. It was taken as snuff in eighteenth century England. Early in the next century the poor smoked pipes while the rich enjoyed cigars. Later in the nineteenth century the rich began to prefer smoking pipes. Officers, returning from the Crimean War, brought cigarettes back with them. English cigarettes were first manufactured in 1856. These made smoking more acceptable for women. The habit became more popular and after the First World War there were more cigarette than pipe smokers.

1 a   How did American Indians use tobacco?
   b   How might tobacco be taken as snuff?
   c   Why might tobacco have first been brought to Europe?
   d   What opposition was there to smoking in the seventeenth century?
   e   Describe how fashions in using tobacco have changed.

2   Scientists have found that cigarette smoke contains carbon monoxide, nicotine and tar. The carbon monoxide gets into your blood, taking the place of oxygen. This can make you short of breath. Nicotine is an addictive drug which makes you rely on cigarettes. It can give you a cough or even bronchitis. Nicotine and carbon monoxide also seem to help along heart disease. Tobacco tar gets trapped in your lungs and damages the tissues. This can lead to lung cancer.
   a   Where else might you find carbon monoxide?
   b   Why can smoking make you very short of breath?
   c   What is it, in cigarettes, that makes people continue to smoke?
   d   Why might teenagers first take up smoking? A sketch will help your answer.
   e   Why has smoking become less popular in recent years?

number of cigarettes smoked daily	increased risk of dying
0	1
5	4
10	8
15	12
20	16
25	20
30	24
35	28
40	32

table 1 Smoking is a risky business

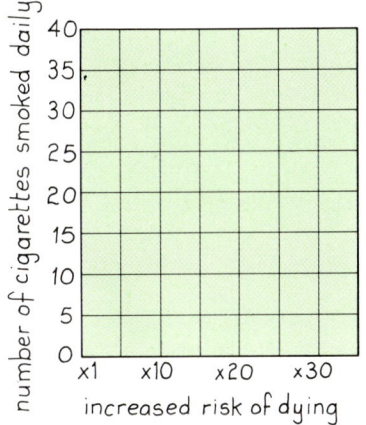

Fig. 42

Grave statistics

3 Table 1 gives details about the estimated risks of dying from lung cancer, related to the number of cigarettes smoked. Plot a graph of the information given. Copy the grid in fig. 42 to start with. What do you conclude from your graph? Have *you* tried smoking?

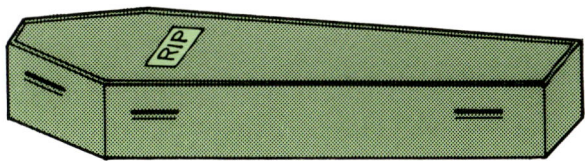

4 A couple have smoked, between them, eighty cigarettes daily for forty years. Estimate the total cost of their smoking at today's prices. How else might they have spent their money?

5 Answer this question in groups of four students. Imagine you are senior citizens. You have been worried by the number of very young children smoking publicly in your town. Plan a campaign to point out the dangers of smoking to the children and their parents. Design a poster to show young children that smoking is dangerous. Give a talk to younger pupils at your school on the subject.

No smoke without fire

1 What are asbestosis and pneumoconiosis? Write about the causes, symptoms and treatment of the diseases. What have they in common?

2 a Write a letter to your local bus company. In your letter argue either for or against smoking on public transport.
b Design an anti-smoking advertisement to fill the space on the bus, fig. 43.

3 It has been said that governments are not too keen to stop people smoking. Why might this be the case?

Fig. 43

Designer bus

Ekopulu

1 A simple system for recycling waste is shown in fig. 44.

In the 'ekopulu' the urine and faeces are kept separate. The urine may be used to irrigate crops. The faeces can be used in grow bags to cultivate vegetables.

a Why is the urine collected in a plastic rather than a metal container? What other materials might be suitable to contain the urine?

b Smell might be a problem for the metal drum containing faeces. What could be done to the drum to cut down the smell?

c Fig. 44 suggests that domestic waste water be added to the urine. List three sources of suitable waste water which might be used.

d Suggest reasons why the urine and faeces should be kept separate.

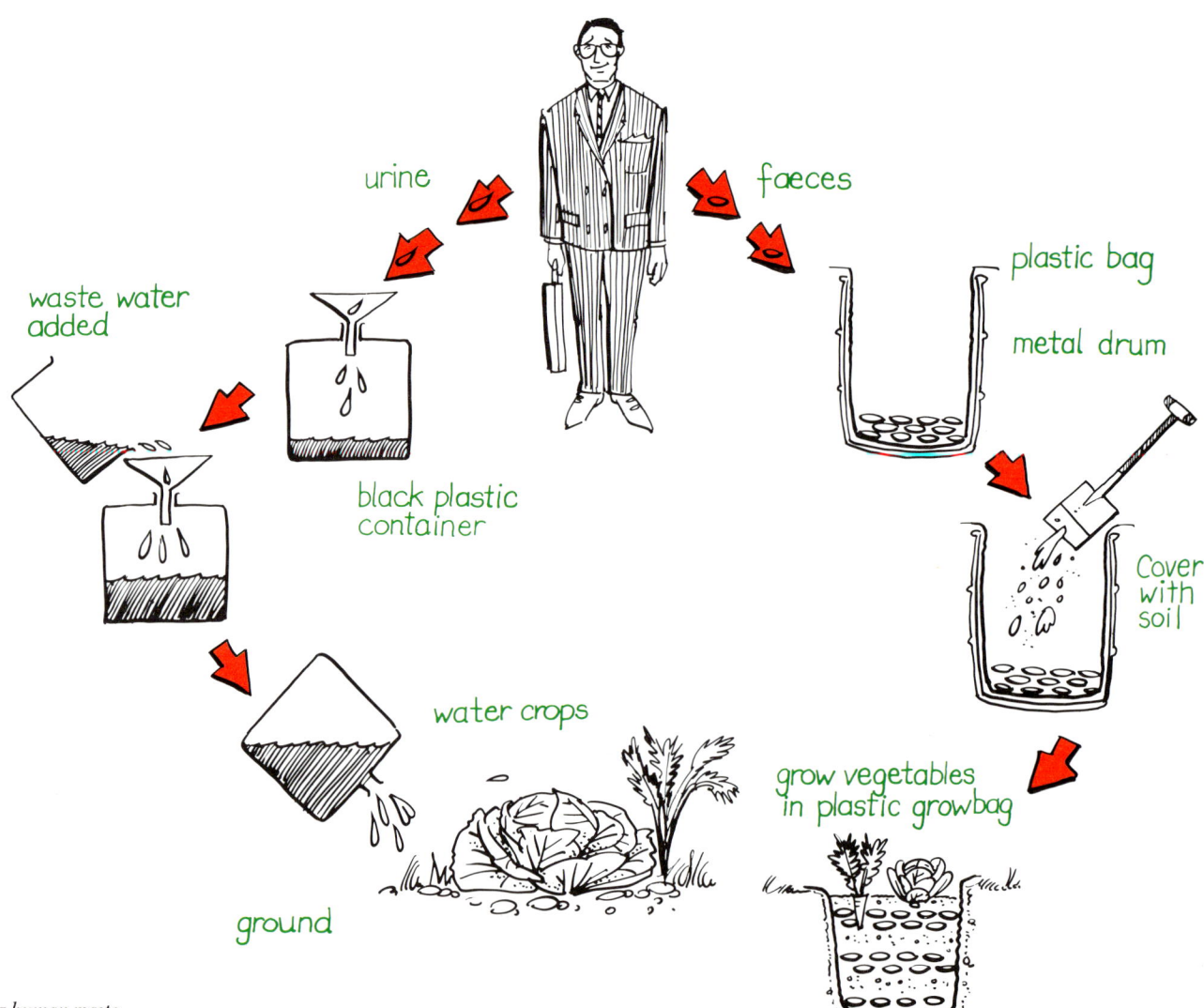

Fig. 44

Recycling human waste

2 a Imagine you wish to set up your own ekopulu. List the materials you need. Estimate the total cost of your system. Which materials are the most expensive?

b Devise and list a simple set of instructions to explain how to set up an ekopulu.

c Draw a flow diagram to illustrate how to use your ekopulu.

3 a Are there any advantages to using grow bags in built-up city areas?

b How might rural areas benefit from the use of the ekopulu?

c List environments where the system might be useful.

4 One human, using an ekopulu, produces annually the equivalent of 25 kg of top class fertiliser. Calculate how much equivalent fertiliser could be produced in the United Kingdom:

a annually

b weekly

c daily.

Assume a population of 50 million.

5 Discuss this question in groups of four students. Imagine you are advertising executives. You work for a company planning to sell ekopulus in the United Kingdom. Your job is to persuade people to buy and use them.

a List the problems you might meet when trying to convince the public to use the ekopulu.

b Where might you best advertise the system?

c Design an advertising poster for the ekopulu.

It's in the bag

1 Bilharzia is a disease which affects the intestine and bladder. It is found in the East, South America and Africa. The disease is caused by a parasite which enters the bloodstream. This parasite has an amazing life cycle.

a Find out about bilharzia and the life cycle of the parasite causing it. Use a diagram to illustrate the parasite's life cycle. How might bilharzia be prevented?

b Claims have been made that by using the ekopulu the occurrence of bilharzia might be reduced. Do you think the claim could be correct? Explain your answer.

2 Information about some of the substances present in urine is given in table 1.

a Which substance forms the greater part of urine?

b What percentage of urine does table 1 not account for?

c Suggest why the concentration of urea in urine is 60 times greater than that in blood plasma.

d State three ways in which your body loses water other than by urinating.

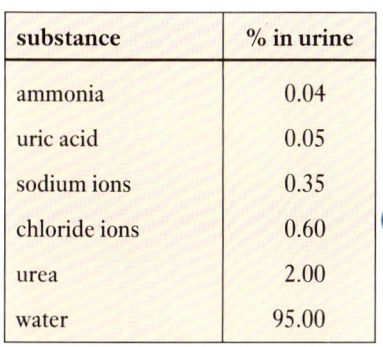

substance	% in urine
ammonia	0.04
uric acid	0.05
sodium ions	0.35
chloride ions	0.60
urea	2.00
water	95.00

table 1 Substances present in urine

Blood ties

Your blood consists of red cells and white cells floating in a pale, yellow fluid called **plasma**. For every white cell you have there are several hundred red ones. 55 per cent of your blood volume is plasma. About 8 per cent of your body mass consists of your circulating blood.

On average a human has about 9000 **white cells** per µl of blood. There are several different kinds of white cell. They defend you against infection by killing germs.

Your **red cells** are shaped like car wheels. They are about 7.5 µm in diameter and 2 µm thick. Red cells are made in the bone marrow. If you are a woman your red cell count is about 4.8 million cells/µl. For men the count is 5.4 million cells/µl. Red cells contain **haemoglobin**. Haemoglobin is a red protein which is at the centre of a red cell. Haemoglobin's job is to carry or transport oxygen, fig. 45, from the lungs to the tissues. Red blood cells also carry carbon dioxide back to the lungs from the tissues.

Fig. 45   Transporting oxygen

The fluid portion of your blood is **plasma**. It is an amazing solution containing many different kinds of molecules and ions. These travel to various parts of the body. Your blood does the following jobs inside you:

- it transports oxygen
- it removes carbon dioxide
- it transports food and removes waste
- it takes away excess heat
- it transports hormones to control vital processes
- it defends you against infection.
  What more could you wish of your blood?

part of blood	colour of part of blood	what it does
	red	
	white	
plasma		

table 1 What's in your blood?

1   Copy and complete table 1. You will find the answers in the passage above.

2 a What is haemoglobin? What does it do?

b Write down your approximate red cell count.

c Summarise the jobs your blood carries out in your body.

3 There are clues in the passage. They tell you what a red blood cell looks like. Write down the clues. Make a scale drawing of a red blood cell. What scale have you used?

4 a There is less oxygen to breathe at high altitudes than at sea-level. Imagine you live at altitude. Would you have more red blood cells than if you lived at sea-level? Explain your answer.

b Why do athletes train at altitude?

5 a How much of your body mass consists of circulating blood?

b Three people are shown in fig. 46. Calculate the mass of blood in each of their bodies.

c About 5 per cent of your body mass consists of plasma. Calculate the mass of plasma in the people shown in fig. 46.

6 a Blood transfusions save many lives. The National Blood Transfusion Service collects blood from donors. In 1980 about two million blood donations were given. However, recently this number has dropped. Suggest reasons for the decrease. How might more people be persuaded to give blood?

b Design a poster to encourage people to give blood. Aim your poster at the 18 to 30 age group. It might help you to consider:

● whole blood, given to road accident victims

● dried plasma, given to burns victims.

Thicker than water

1 a Haemoglobin is found in the blood of all mammals. What does this suggest about their ancestors?

b White blood cells defend you against infection. Find out and describe how white cells kill bacteria.

2 a Leukaemia is a disease. It is sometimes called cancer of the blood. Find out and write about the causes and treatment of this disease.

b Coronary thrombosis is a disease. It is the most common cause of sudden death. Find out and write about coronary thrombosis.

3 a Carry out a survey of your fellow students. Find out how many would be prepared to donate blood.

b Write a letter to the National Blood Transfusion Service. In your letter ask for information on donating blood.

75 kg.

15 kg

50 kg

Fig. 46

Blood relationships

Reaction driving

Jean and Jenny Begg were twins. Their mum usually drove them to school. Mrs Jones was the girls' year head at school. One morning she stepped out in front of the twins' car without looking. Mrs Begg had to make an emergency stop. Luckily she braked in time. Mrs Begg was most annoyed and gave the teacher a stern lecture on road safety. The girls were proud of their mum's quick reaction. They wondered if all drivers would react so quickly. Jean and Jenny designed an experiment to find out. They used fellow students as guinea-pig drivers. Jenny's unfinished notes on the experiment are shown in fig. 47.

Fig. 47

Jenny's unfinished notes

1 a What does the squeaky toy represent in their experiment?
  b With which foot does the driver press down on the squeaky toy?
  c The twins used a red light in their investigation. Suggest why they chose this colour.

2 a Copy and complete Jenny's bar chart. Use graph paper.
  b How many drivers did the twins test?
  c Which reaction time was the most common? What else did they find out?

3 a List all the equipment they might have used. It's not all shown in fig. 47!
  b When did the twins start to time the driver? When did they stop?
  c Where might Jean and Jenny have stood while testing the driver? Explain your answer.
  d Briefly describe how you would have carried out their investigation.
  e Suggest a substitute for the squeaky toy in the experiment.
Use a diagram to explain how your substitute works.

4 Fig. 48 gives information about Mrs Begg's car brakes. It shows the overall stopping distance at different speeds.

a What happens to the overall stopping distance as the car travels faster?

b Learner drivers must study the Highway Code. This suggests that the overall stopping distance consists of two parts. These are *thinking distance* and *braking distance*. Explain the difference between them.

c Suggest one factor that might affect braking distance.

d Describe one factor that could affect thinking distance.

e Copy and complete the table shown in fig. 48. Use the information given in the graph.

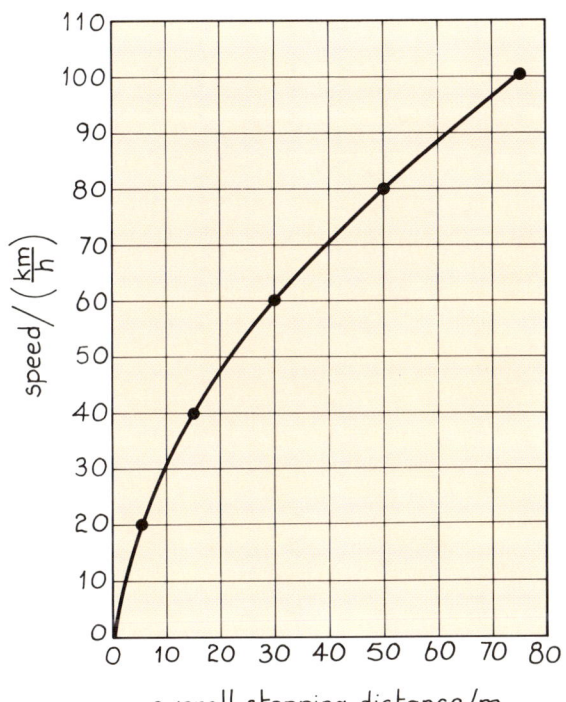

speed/$\left(\frac{km}{h}\right)$	overall stopping distance/m
20	
40	
60	
80	
100	

Fig. 48

Stopping Mrs. Begg's car

5 Ambulances often go faster than the speed limit. Other road users need to be warned of their approach. Design an ambulance sign that can be seen clearly by drivers in front. Test your design using a mirror.

Driving you mad

1 Your pupils control the amount of light entering your eyes. How does pupil diameter change with the brightness of light? Fig. 49 shows the results of an experiment to find out.

a Use a mirror to help you sketch the pupil of your eye.

b Some drivers wear sunglasses at night. Suggest why this is dangerous.

c Design a simple experiment to test the results shown in fig. 49.

d How quickly do your pupils react to changes in brightness? Should you drive off quickly after coming out of a brightly lit building? Explain your answer.

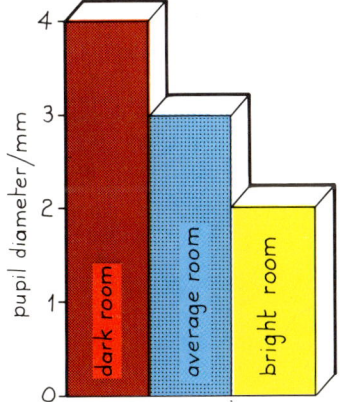

Fig. 49

Pupil size and brightness!

Life cycles

Salmon are examples of fish. Fish live in water and have no limbs. Fish have a streamlined shape to help them move through the water. They have powerful muscles and their bodies are covered in scales. Their fins help them to swim. The tail fin propels them forwards. They are kept upright by their anal, dorsal and ventral fins. The pelvic and pectoral fins help them to steer and to stop.

Salmon have an amazing life cycle. The following passage gives you some clues to their life history.

During the summer the . . . . salmon lays her eggs. The eggs are fertilised by the male and hatch into . . . . at the end of winter. After one year the young fish is about . . . . long. During its second year it grows to . . . . in length. In the spring of its third year the salmon . . . . , via the . . . . , to the sea where it lives for several . . . . . The fish then returns, up the same river, ready to . . . . .

1 Copy out the passage on the life cycle of the salmon. Fill in the blanks with words from the box.

| 15 cm | female | spawn | years |
| river | fish | migrates | 10 cm |

name	age	where it is
alevin	to one year	river
parr	one year	river
smolt	two years	river
grilse	to four years	sea
salmon	four years	returns to river
kelt	after four years	returns to sea

table 1 History of a salmon

2 Study fig. 50 and table 1.
Either
a draw a flow diagram to summarise the life cycle of a salmon;
or
b write and illustrate a story about the salmon's life cycle. What do you think happens to the salmon after it has spawned?

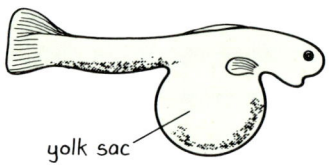

yolk sac

Fig. 50

Baby and adult salmon, not to scale

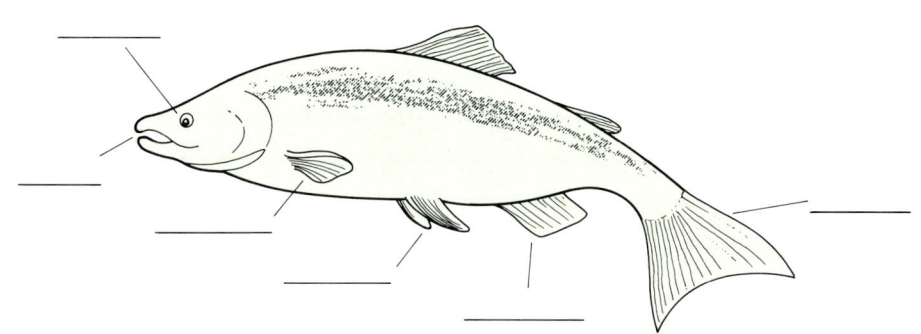

3 Copy and label the adult salmon drawn in fig. 50. Choose your labels from the box.

tail fin	ventral fin	pectoral fin
mouth	eye	pelvic fin

Fishy finishers

1 The maximum speed that a fish can travel might depend on *the length of the fish*. Look at table 2 which lists fish in order of increasing length. It gives information which can be used to work out the maximum speed of the different species.

a Copy and complete table 2.

b Which fish can swim the fastest? Which fish is the slowest swimmer? How, do you think, does the length of a fish determine its maximum speed?

c Which species shown in table 2 is a mammal?

species	distance travelled/cm	time taken/s	maximum speed/cm s
goby	100	4.00	25
stickleback	100	1.40	
goldfish	100	0.70	
herring	100	0.60	
rainbow trout	100	0.60	
pike	100	0.50	
dolphin	1000	0.55	
blue-fin tunny	1000	0.50	

table 2 Maximum speeds of fish

2 a Find out how fish breathe. Draw a labelled diagram to show how fish use their gills to breathe.

b Find out how fish swim. Draw a labelled diagram to show how fish move forwards by moving their tails from side to side.

c Do dolphins swim by moving their tails from side to side? Draw a labelled diagram of a dolphin.

Baby growth

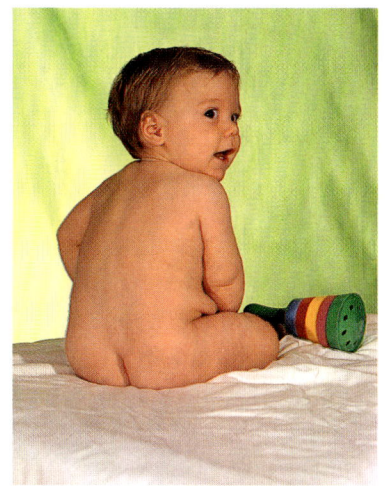

Ann

Beth

Claire

Do you know what your mass was at birth? Family doctors and health visitors find it useful to know how a baby grows in its early years. They keep a record of the mass gained by a baby as it grows. These records allow doctors to check that a baby is both feeding well and thriving.

Look carefully at the photos and table 1. Ann, Beth and Claire were born on the same day at the same hospital. Their parents measured and noted the girls' masses up to the age of two years. In Britain, about one in twenty babies is smaller than Claire. Beth is of average size. About one in twenty babies is larger than Ann.

age/months	Ann's mass/kg	Beth's mass/kg	Claire's mass/kg
0	4.5	3.5	2.5
4	8.2	6.5	5.0
8	10.5	8.5	7.0
12	12.3	10.0	8.1
16	13.2	10.9	9.0
20	14.2	11.8	9.5
24	15.0	12.5	10.0

table 1 Growing girls

1 a Which baby was heaviest at birth?
b Which baby was lightest at birth?
c Find the average mass at birth of the three babies.

2 Using the same axes draw a graph to compare the growth of the three babies. Plot *mass* against *age* for each baby as in fig. 51.

Fig. 51

Growing babies

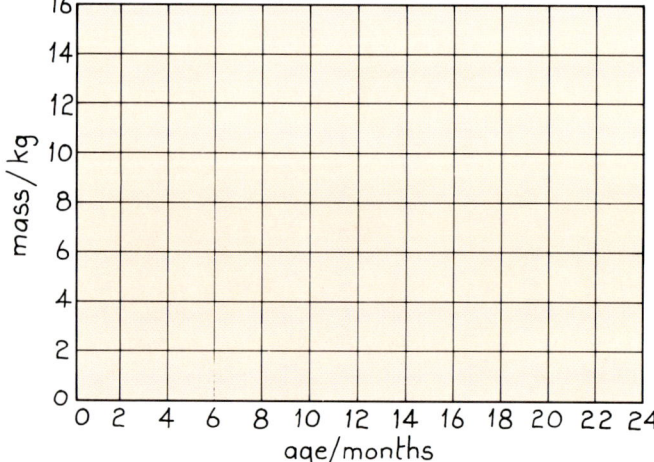

3 From your baby graph answer the following questions.
  a At what age was Ann 5.5 kg?
  b At what age was Beth 5.5 kg?
  c At what age was Claire 5.5 kg?
  d How old was each of the girls when she was 9.0 kg?

4 a Write down your own mass at birth.
  b Convert your mass at birth to kilograms.
    1 pound = 0.45 kilogram
    1 ounce = 0.028 kilogram

5 a What might Ann's parents have used to measure her mass?
  b What advice would you give to Ann's parents to make sure that the masses they measured were accurate?
  c Draw a flow diagram to show how Ann's parents might have measured her mass.

6 Fig. 52 shows the growth curves for pupils at Lurchfield School.
  a When were the girls heavier than the boys?
  b Describe the similarities and differences between the two curves.
  c How do you expect the masses of the boys and girls to change after the age of sixteen? Explain your answer.

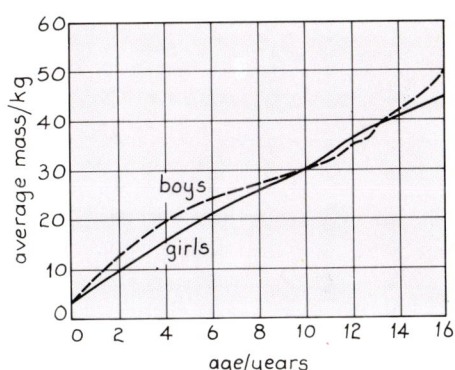

Fig. 52

Lurchfield babies

Grow on from here

1 Why are there wide variations in mass at birth?
  a Discuss your ideas in groups of four students.
  b Draw a poster to illustrate your group's ideas.

2 Midwives and health visitors do very important jobs in the community. Find out and write about what they do.

3 Table 2 gives information about Beth's height.
  a Plot a graph to show how Beth's height changed with age.
  b When was her rate of growth the greatest?
  c When was her rate of growth the least?
  d Do you think Beth will grow taller after sixteen years of age?

age/years	height/cm
0	50
4	100
8	130
12	150
16	160

table 2 Beth's height

Populating Rar

Rar, fig. 53, is an imaginary planet in the constellation Orion. Its land is divided into three continents. These are *Attico*, *Avknot* and *Avland*. Rarons inhabit the planet. There is little contact between Rarons from different continents.

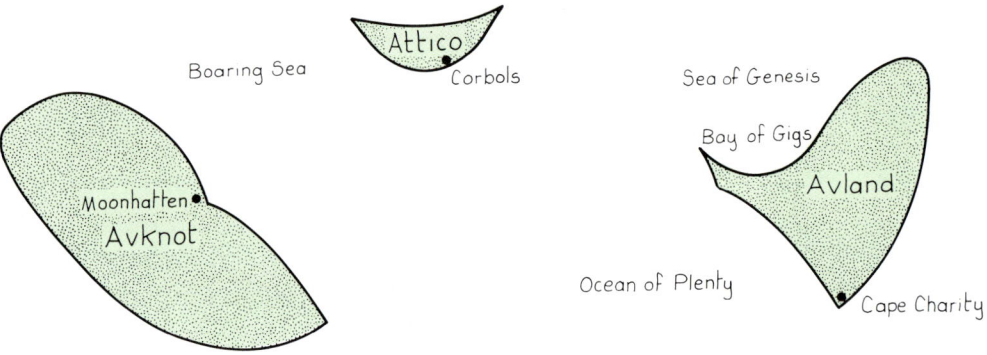

Fig. 53

Planet Rar

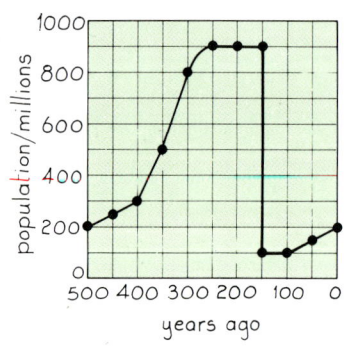

Fig. 54

Rar's population

1 a Scientists on Rar believe their three continents were joined together 100 million years ago. Sketch Rar's landmass as it might have appeared then.
b Suggest why the continents might be moving.

2 Rar's total population, over the past 500 years, is shown in fig. 54.
a Write down the population of Rar:

- 500 years ago

- 250 years ago

- 100 years ago

- now.

b Suggest possible reasons for the sharp drop in population.
c Estimate Rar's population in 100 years time.

3 Life expectancy varies on Rar. Details of life expectancy, on the three continents, are given in fig. 55.
a Explain the term *life expectancy*.
b On which continent is life expectancy the greatest? Where is it the least?

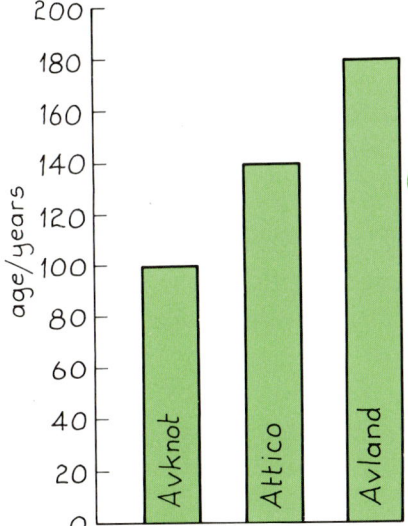

Fig. 55

Life expectancy on Rar

Fig. 56

Avland's population pyramid

c Estimate the age range of *adolescents* on Attico. Explain your answer.

4 Fig. 56 is a population pyramid. It gives information about Avland's present population. Table 1 describes Attico's present population.

a What percentage of Avland females are aged between 140 and 159?

b What percentage of Attico females are aged between 140 and 159?

c Which sex lives longer on Rar? Explain your answer.

d Use the information in table 1 to plot a population pyramid for Attico. Use graph paper.

e List the similarities between the population pyramids of the two continents.

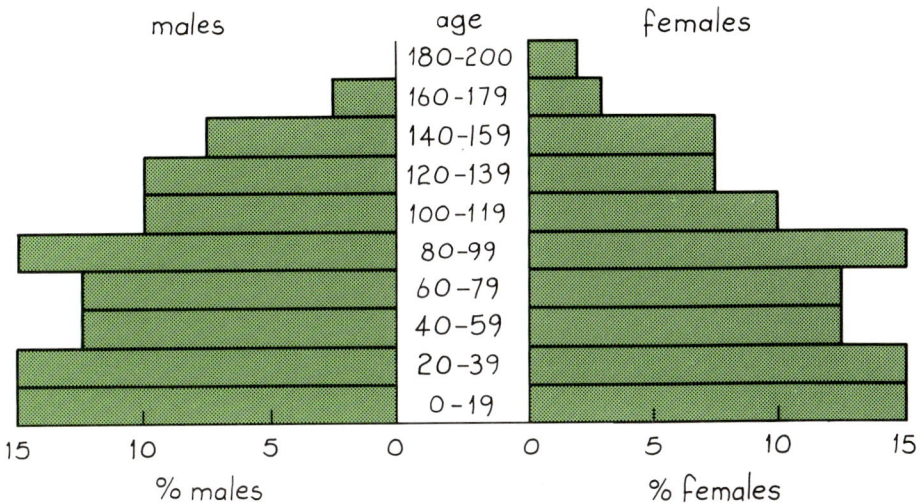

5 Answer this question in groups of three students. Why might life expectancy vary on Rar's three continents?

a List factors which might affect life expectancy on Rar.

b Draw a poster to illustrate how Avknot's life expectancy might be increased. Use your imagination!

6 Do you think Rarons from the three continents might resemble one another? Explain your answer. What factors could change a Raron's appearance? Draw sketches to explain your answer.

Rar-ing to go

1 Scientists on Rar investigated how the number of rarbugs increased with time. Their results are shown in table 2.

a Plot a graph of their results.

b From your graph estimate the number of rarbugs after 25 minutes.

c Rarbugs are the size of pinheads. Suggest how the scientists might have estimated their number.

2 Imagine yourself as a Raron visiting Earth for the first time. Write a brief passage describing your impressions of Earth and its inhabitants. How would the people on Earth react to your visit?

ages/years	% males	% females
0–19	25.0	25.0
20–39	22.5	17.5
40–59	12.5	12.5
60–79	12.5	12.5
80–99	15.0	15.0
100–119	10.0	10.0
120–139	2.5	5.0
140–159	0.0	2.5

table 1 Attico's population by age

time/min	number of rarbugs
0	100
10	200
20	400
30	800
40	1600

table 2 Rarbugs

Sunflowers

part of sunflower	use
stalks	
	yellow dye
seeds	
	fodder
	oil

table 1 Using sunflowers

1 Samuel de Champlain was a French explorer. Early in the seventeenth century he visited the Great Lakes of North America. On the eastern shore of Lake Huron he found American Indians, cultivating large-headed sunflowers. They used the stalks as textile fibres and the leaves for fodder. The flowers produced a yellow dye. The seeds were used as food and also as a source of oil. Sunflowers were introduced into Britain in 1596. Today they are grown commercially on a large scale in many parts of the world.

    a   Copy and complete table 1.

    b   Can you write down any other modern uses of sunflowers?

2 Growing sunflowers involves lots of hard work. Re-arrange the following jumbled up stages. When re-arranged they will tell you how to grow and measure a sunflower:

- measure the final height with a metre-rule

- as the plant grows taller support it with a stick

- give the plants fertiliser from time to time

- plant the seeds in seed trays containing potting compost

- when large enough transplant the germinated seeds into the garden

- water the seeds regularly , don't waterlog the compost

- after transplanting keep the soil moist, not too wet.

3 Students at Midshire School organised a sponsored sunflower competition. They were raising money to buy a new school minibus. Table 2 gives information about the winning sunflower.

    a   Plot a graph to show its growth.

    b   From your graph find the height of the sunflower at five weeks.

    c   When was the rate of growth of the sunflower the least?

4 Answer this question in groups of four students. Discuss and design a sponsored sunflower competition to raise funds for your school. Some points you might wish to consider are:

- when will the competition take place?

- what is to be sponsored, e.g. height, weight, flower size etc.?

- who will judge the competition?

- sunflowers grow very tall!

height/cm	age/days
25	10
80	20
150	30
220	40
280	50
320	60
330	70
335	80
336	90

table 2 The winning sunflower

Field of sunflowers in Provence, France

a   Draw a series of cartoons to explain to competitors how to grow their sunflowers.

b   Design a sponsorship form for the competition.

c   Design a results table for the competitors to use.

d   Write a set of instructions to explain how the sunflowers should be measured accurately. A diagram would also be helpful.

5   Believe it or not, the stumps of giant sequoia trees have been used as dance floors!

a   Estimate the cross-sectional area of a giant Californian sequoia. Its radius, r, is 3 metres.

$$area\ of\ circle = \pi r^2$$
$$\pi = 3.14$$

b   Find the volume of the same tree. It is 70 metres high.

$$volume = area \times height$$

c   Estimate the mass of the sequoia. Its density is $500\,kg/m^3$.

$$mass = density \times volume$$

d   Is your estimate of the tree's mass reasonable? Explain your answer.

Further fodder

1   Write a short article. In your article describe either the short term or long term effects of a world shortage of wood. Suggest how we can ensure a good supply of wood for future generations.

Cyperus papyrus from the Nile

2   Papyrus was used in ancient Egypt until the fourth century AD as a source of paper-like material. The plant is now extinct in lower Egypt but still grows around the Upper Nile. The word 'paper' comes from papyrus. Find out how the early Egyptians made papyrus paper. Draw a flow diagram to describe its manufacture.

3   The school gardener has advised you that sunflower seeds need water, warmth and oxygen to germinate. Design a series of simple experiments to find out if she is correct. Don't forget to include controls in your experiments.

4   Fruits are made by plants to help them disperse seeds.

a   What is the difference between pollination and dispersal?

b   Find out how dandelion and sycamore seeds are dispersed.

c   Design a seed that can be dispersed by the wind! Draw a labelled diagram of your seed.

Ovulation

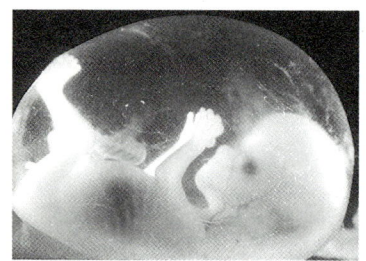

Life in the womb

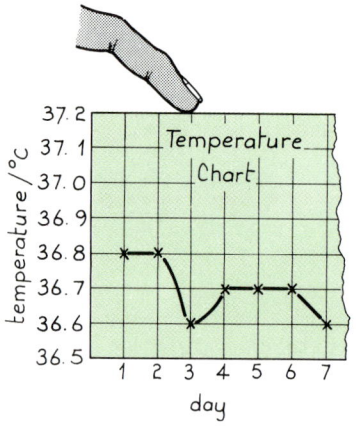

Fig. 57

Recording temperature

Sue and Bob have been married for five years. For the past year they had been trying, without success, to start a family. Sue suggested to Bob that they visit the family doctor with their problem.

Dr May explained to the couple that Sue was most likely to conceive just after the time she ovulates or releases an egg from her ovary. The egg must be fertilised by Bob's sperm within twenty-four hours if she is to conceive. The doctor explained that most women ovulate about 14 days before their next period.

Dr May suggested that Sue should keep a record of her temperature in the form of a chart, fig. 57. The temperature chart would show when Sue ovulated. It might also show if Sue was not ovulating.

Sue was given an accurate thermometer called a fertility thermometer. Dr May explained that Sue should take her own temperature daily, first thing in the morning, starting on the first day of her period.

Sue asked how the chart would show when she ovulated. The doctor explained that about 14 days before her period her temperature would go up. This temperature rise would mean she had just ovulated and that the most likely time to conceive had just passed. However, an egg can live in the fallopian tube for about twenty-four hours. So Sue could still conceive if she had intercourse just after ovulation.

Sue followed the doctor's advice and kept a daily temperature chart. Twelve months later Sue gave birth to Jenna, a bouncing baby girl.

1 Copy and complete the following sentences. Choose your answers from the words in the box.

conceive	fallopian	fertilisation	ovulate

a To . . . . means to become pregnant.
b On being released from one of the ovaries the ripe egg travels down the . . . . tube.
c To . . . . means to release an egg.
d When a male sperm unites with a female egg . . . . has occurred.

2 Draw and label fig. 58. Choose your labels from the words in the box.

egg	fallopian	ovary	vagina

day	Sue's temperature/°C
1	36.8
2	36.8
3	36.6
4	36.7
5	36.7
6	36.7
7	36.6
8	36.8
9	36.6
10	36.7
11	36.6
12	36.7
13	36.7
14	36.7
15	37.0
16	37.0
17	37.0
18	37.1
19	37.1
20	37.1
21	37.0
22	37.1
23	37.1
24	37.1
25	37.0
26	37.0
27	37.0
28	36.9

table 1 Sue's temperature record

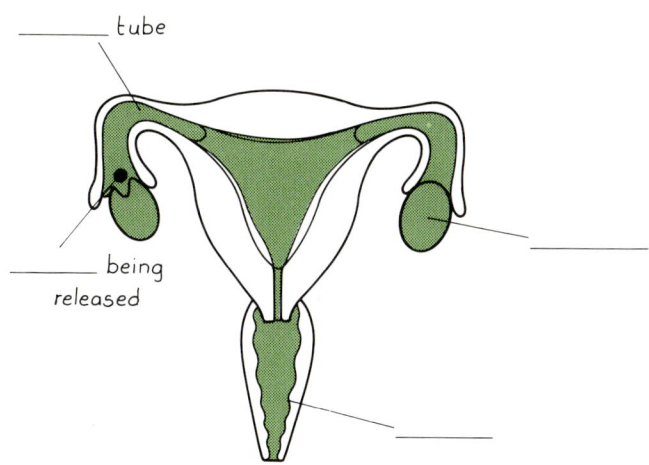

Fig. 58

Female reproductive system

3   Why did Dr May suggest that Sue should keep a daily temperature chart?

4   Look carefully at table 1. This is a record of Sue's daily temperature over a monthly cycle.
   a   Plot a graph of *Sue's temperature* against *day*. Join the points together. This is Sue's temperature chart.
   b   Look at Sue's temperature chart. Over which two days is there the sharpest rise in temperature?
   c   When did Sue ovulate?

5   Had Sue not conceived after another twelve months what might have happened next?
   a   Discuss your ideas in groups of four students.
   b   Write a short summary of your group's ideas.

Read on

1   **a**   Why did Dr May give Sue an *accurate* fertility thermometer to measure her daily temperature?
   b   Why did Sue have to take her temperature first thing in the morning?

2   Answer this question in groups of four students.

> Who needs men? One adult male will produce over 300 million sperms every day. We only need one man in the world for the race to survive.

Talk about the point of view shown in the box. Write a short article for *The Daily Nag* newspaper discussing the point of view.

3   Write a short poem about life in the womb.

A dog's life

1   Look at fig. 59. It suggests when some animals may have been first domesticated.
   a   Explain the term *domestication*.
   b   When were reindeer domesticated?
   c   When were yaks first domesticated? Give your answer in years BC.
   d   Which animals, in fig. 59, are used for meat by man?
   e   List the animals, in fig. 59, used for transport.
   f   Did man first domesticate animals for transport or for meat?
   Explain your answer.

2   Put the information shown in fig. 59 into the form of a table. Your headings might be *type of animal* and *when domesticated*.

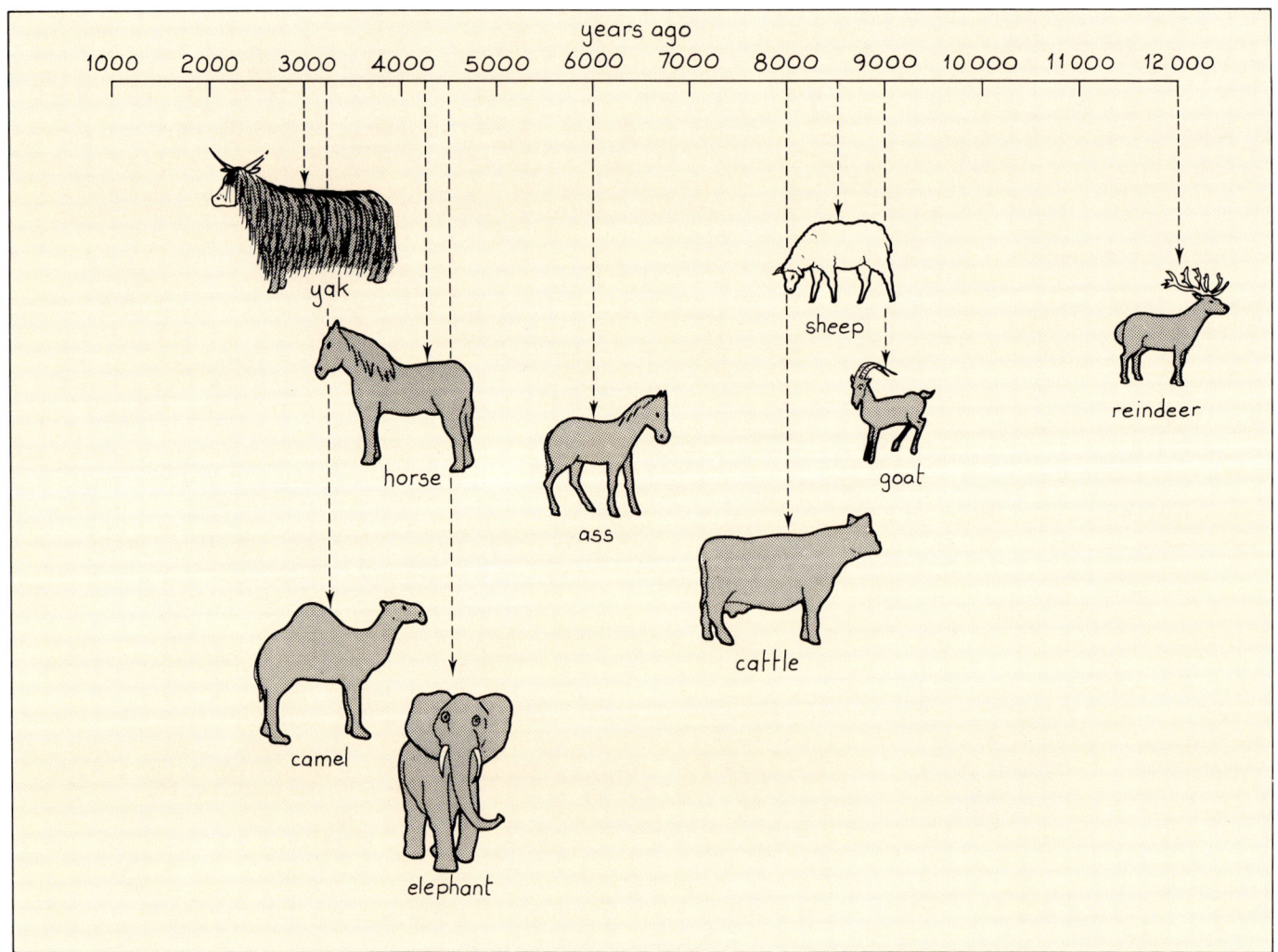

Fig. 59   *Domesticating animals*

3 Packs of wild dogs hung around human campsites as long ago as 15 000 BC. There they may have fought for scraps. Later dogs may have guarded the campsites against other animals. They then helped man to hunt and so earned their keep. Later man began to breed dogs for specific jobs. Over 5000 years ago small terriers and greyhound types were used by the ancient Egyptians. Mastiff dogs helped the ancient Phoenicians to hunt lions. The Phoenicians were great sailors who brought the mastiff to Britain.

    a   List four different breeds of dog. What work were they bred for?

    b   Explain why the dog became man's best friend.

    c   It has been suggested that dogs may have evolved from wolves. List the similarities and differences between them.

4 a   Bulldogs and terriers are different breeds of dog. Dogs from each breed were carefully chosen and mated together. Eventually a new breed was created known as the Staffordshire bull terrier. Why should people wish to produce a new breed of dog? Explain what is meant by the dog's pedigree?

    b   Fashions in owning dogs change. How might fashions in owning dogs affect their population?

5 Answer this question in groups of four students. There is a great variety of dogs in existence today. Discuss and list the factors which may have caused this wide variation. Design a poster to summarise your group's ideas.

Domestic science

1 a   Which breeds are used as guide dogs for the blind?

    b   Find out how guide dogs are trained. What special qualities are required of a guide dog?

2 Cats occur in many varieties. The word 'puss' comes from the ancient Egyptian 'Pashta', meaning the cat goddess. Cats were worshipped in ancient Egypt and were even embalmed and laid to rest in tombs!

    a   Of what use might cats have been to the Egyptians? Suggest reasons why they may have worshipped their cats.

    b   Why were cats, in Britain, often persecuted in the Middle Ages?

    c   Suggest reasons for the wide variety of cats in the world today.

3 The ancient Phoenicians were great travellers and traders who were clever enough to devise an alphabet of 22 symbols. They even visited Britain to buy Cornish tin. Find out:

- where their homeland was

- the names of their two large cities

- what contributions they made to world industry.

Catering for different tastes

Matey moorhens

Moorhens might be described as little jerks! They jerk their heads while swimming and their tails while walking. The moorhen is a common bird which can be found on inland water. It is about 30 cm in length and is resident in Britain. The female lays six to twelve eggs between April and July. Moorhens may eat seeds, berries, insects, worms and water plants.

The males do most of the incubating of eggs. There are usually lots of males available. The females compete with each other for the best quality males. It is the female that starts the courtship. The hens will fight if a courting pair is approached by a female. Usually the fatter hen wins. Strangely the fatter, winning hens seem to end up with the smaller males. Perhaps the best quality males are the small ones! The larger males have a smaller proportion of bodyweight fat than the smaller ones. This means the smaller males have greater food reserves and can sit tight on the eggs for longer periods. Could this be the reason why the hens prefer the smaller males?

1 a   Why are moorhens called little jerks in the passage above?
   b   List four places where moorhens might be found.
   c   What materials might a moorhen use to build its nest?
   d   Does the moorhen migrate from Britain?

2 a   Who incubates the moorhen's eggs?
   b   Why do females fight to keep their males? Which female usually wins?
   c   What type of mate does a large female usually select?
   d   Explain why the large hen might choose such a mate.

3 Look at table 1. It gives details about lemurs, chimpanzees and man.
   a   Put the information in the form of a bar chart or charts.
   b   Write a short summary of the details given in table 1. How do the phases of chimpanzees compare to those of humans?

	lemur	chimp	man
pregnancy/days	126	238	266
infant/years	0.5	3	6
juvenile/years	2	7	14
adult/years	11	30	55

table 1 Phases in growing up

4 Answer this question in groups of four students. Discuss and write down your answers to the following questions:

● Was there a time before which no life existed?

● How did the first species come about?

● How might other species have followed the earlier ones?

● Where did humans come from?

● Why do humans have different coloured skins?

Design a poster to illustrate your group's ideas.

Selected questions

1 Charles Robert Darwin lived between 1809 and 1882. He was one of Britain's most controversial scientists. Find out and write about Darwin's life. How did Darwin's work change man's outlook on his origins?

2 There have been many hoaxes throughout the history of science. One such hoax was known as the Piltdown Man. Find out what the Piltdown Man hoax was about. What other hoaxes have there been in science?

3 Suggest reasons for the following observations:

● giraffes have long necks

● elephants have long trunks

● polar bears have white fur

● gorillas have no tails.

4 The volume of a chimpanzee's brain is 400 cm^3 whereas that of a gorilla is 500 cm^3. Do you expect a human's brain, to be smaller or larger? Explain your answer. How are humans different from other apes?

African elephant

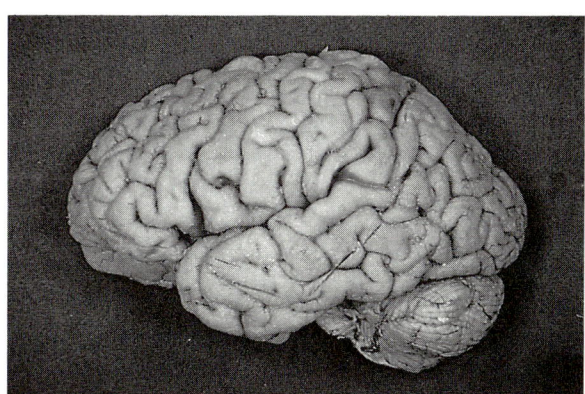

Brainbox

Cholera detective

Cholera seems to have come from Bengal, India. It spread through the world during the nineteenth century, killing many people. In Hamburg 17 000 people came down with cholera in a single epidemic. Over half the victims died. In 1854 a serious outbreak occurred in London. It started on 19 August and was to last 42 days. Over 600 people died during the outbreak.

The disease has an incubation period of two to five days, after which symptoms appear. The illness has three stages. During the short first stage the patient has a mild case of the runs and wants to vomit. The victim is very thirsty and has severe muscle cramps due to a lack of salts. In the second stage the patient is in a state of collapse. Blood pressure is very low and the cramps become agonisingly painful. It is during the second stage that the patient may die. Survivors continue to the third, recovery stage.

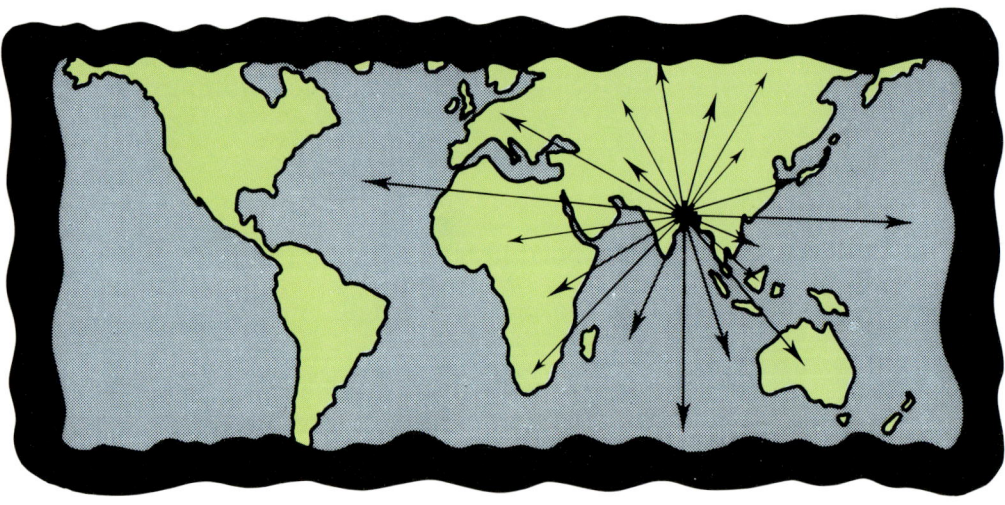

Fig. 60

Spreading cholera

1 a From where did cholera originate?
   b What is meant by the incubation period of a disease?
   c Why might a cholera victim be thirsty?
   d Why does a cholera sufferer experience painful cramps?
   e Explain why cholera victims should have their bedding sterilised.

2 Imagine you are a cholera detective. You are hoping to find the source of the London outbreak. You suspect the source has something to do with a water pump in the area. Fig. 61 describes where the deaths occurred. Find the suspect pump.
   a Which pump did you choose? Explain your choice.
   b You have now found the suspect pump. How might you stop the epidemic?

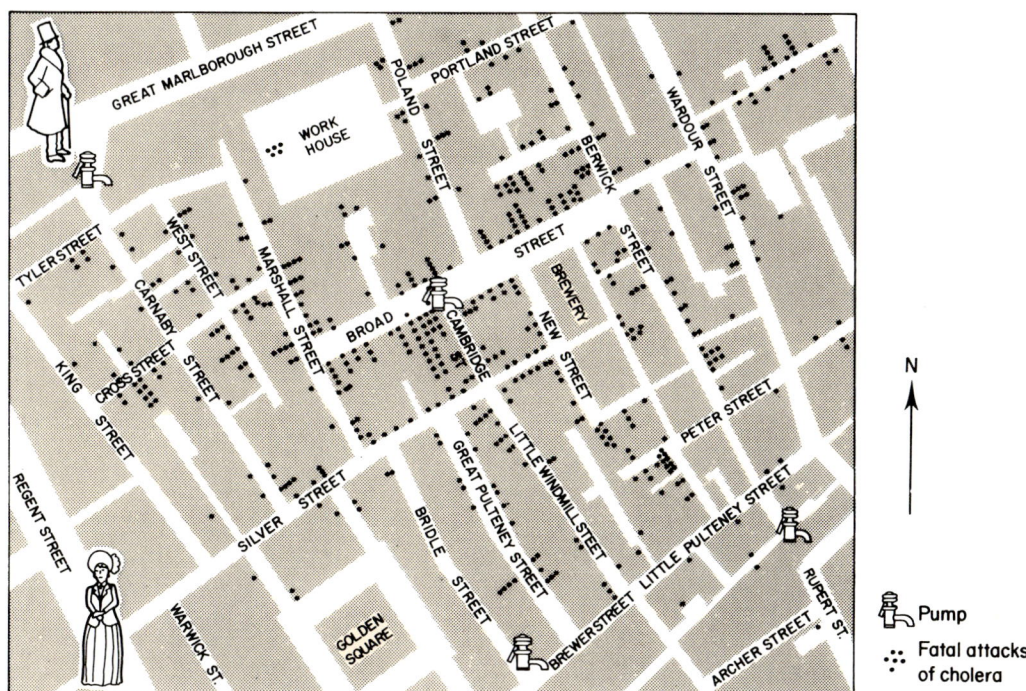

Fig. 61

Cholera deaths in London, 1854

Pump

Fatal attacks of cholera

3 Answer this question in groups of four students. Imagine you are hospital workers. A suspected case of cholera has just been admitted to your hospital. You want to find out about the patient.

a What information might you collect directly from the patient?

b Design a simple questionnaire to help you find out about the patient. Why do you need this information?

DISEASE KILLS BRITONS

Three Britons, all staying in the same hotel in Benidorm, have died from Legionnaires' disease.

Fig. 62

False headline

The state of Britain's beaches?

Beachcombers

1 Sometimes newspapers get their facts wrong. In 1981 a false outbreak of Legionnaires' disease was reported, fig. 62. How might this have happened? What effect do such headlines have on tourists? Write a follow-up article to correct the newspaper's mistake.

2 Many people are worried about the state of Britain's beaches. Why are they worried? What can be done to tidy up our beaches? Draw a poster showing the possible dangers on the beach.

3 Modern treatment of cholera may involve giving a saline solution intravenously to the patient. Up to ten litres of saline solution may be given per day. Antibiotics may also be given to rid the intestines of cholera organisms.

a Explain the term *intravenous*.

b What is a saline solution? Why is it given?

c Why has cholera disappeared from Europe? Under what conditions might it return?

Answers

Pasteurisation
4 (c) 1.8×10^8 l
5 (c) £9 276 966
Milk matters
2 (a) 1.644×10^7 l (b) 0.33 l

Natural cycles
2 (c) 99.9901%

River of death
5 (a) 25 000 m³ (b) 25 m³ (c) 10 d

Weather
2 (c) 94 500 Pa
3 (b) 490 mm (c) 3.5 °C

Sounds medical
1 (b) 70 db
Sounds awful
2 (a) 320 m/s (b) 3200 m

DDT
Creepy crawlers
3 (b) plankton 5.12 ppm fat of fish 10 ppm
        fat of grebes 1600 ppm

Rain, rain go away
2 10^9 kg

It's a gas
3 (b) 6%

Rumbling old Earth
2 (a) 6370 km

Titan
Titanic struggle
2 (b) radius 2.56×10^6 m
    (c) volume 7.03×10^{19} m³
    (e) mass 1.34×10^{23} kg

Lunar-cy
Lunar boosters
2 (b) 0.55 d

Ekopulu
4 (a) 1.25×10^9 kg (b) 24.04×10^6 kg
    (c) 3.42×10^6 kg
It's in the bag
2 (b) 1.96%

Blood ties
5 (b) woman 4.0 kg toddler 1.2 kg man 6.0 kg
    (c) woman 2.5 kg toddler 0.75 kg man 3.75 kg

Reaction driving
2 (b) 31 (c) 0.6 s

Life cycles
Fishy Finishers
1 (a) in cm/s: stickleback 71.4 goldfish 142.9
                  herring 166.7 rainbow trout 166.7
                  pike 200 dolphin 1818.2
                  blue-fin tunny
                  2000

Baby growth
1 (c) 3.5 kg
3 (a) in months: 1.0 (b) 2.5 (c) 4.9
    (d) in months: Ann 5.4 Beth 9.5 Claire 16

Populating Rar
Ra-ring to go
1 (b) 580

Sunflowers
3 (b) 187 cm
5 (a) 28.26 m² (b) 1978.2 m³ (c) 989 100 kg

Ovulation
4 (b) days 14 and 15 (c) day 14